for Ida Faré

Sensefulness
New paradigms for Spatial Design

Nuovi paradigmi per lo Spatial Design

Anna Barbara

© 2018 Postmedia Srl, Milano

Translated by Steve Piccolo

www.postmediabooks.it
ISBN 9788874902378

Sensefulness

New paradigms for Spatial Design
Nuovi paradigmi per lo Spatial Design

Anna Barbara

postmedia•books

Foreword

This book sets out to explore what is happening in the world due to the digital revolution and its repercussions on the design of spaces. For people like me who conduct research and teach in universities, but also design things around the world, there are many aspects that are starting to no longer make sense. So I have decided to stop for a moment and try to at least put the questions that crop up on the pages of my notes into order, jotted down during reviews of the work of my students, listening to younger members of my staff, during meetings with clients, or while I browse through the news of an increasingly compressed and often more iniquitous world.

We are living in a complicated period. We thought we had gained certain rights on a permanent basis, and instead we are leaving young people with the need to defend them, tooth and nail. This harsh and fascinating period demands the sharing of scenarios. Those who think they can find shelter behind some wall will live under siege, just as those who seek refuge in the web will find themselves in a state of alienation.

Outside this racing world there are those who have lagged behind, the slow ones, those who have missed a turn on this dizzying carousel, who have been ejected from the world of work and can no longer find their way back in. There are the senior citizens who no longer understand what language we are speaking, people standing in line at soup kitchens while others work off calories at the gym. There are those who put up resistance, activists or populists. There are the poor, the refugees, the shipwrecked, the desperate ones in the middle of a sea, on the long march to a borderline, on the sidewalk outside my house.

We are all clutching a cellphone, to post, to work, to search for something; to ask rescuers for help, to send our location and get saved, to alert the family as survivors without any other passport.

I have never believed, even for a minute, that the world I am writing about is the only world that can exist.

I think about how the lectures I give to my students must sound, students from all over the world who I want to teach to be designers, not just to do design.

Part 1

The design of spaces

I think about how the lectures I give to my students must sound, students from all over the world who I want to teach to be designers, not just to do design.

I want to help them to understand how to be designers in a world in transformation for which my colleagues and I are at times not even capable of providing convincing interpretations.

With the same respect and responsibility, I have thought about my clients, who seem to have gone through a genetic mutation since I began to design, with whom I feel the need to reformulate a series of paradigms and values that have an impact on both of us.

What has happened to spaces with the advent of the digital era also has to do with design. All the universities on the planet, all the design studios and places of reflection on design, are asking themselves about this.

It is as if the transformations of "places" had happened more rapidly than the thinking that was supposed to design them, and we find ourselves living in "spaces" that are often designed with the habitat logics of the last century.

Yet the digital revolution has profoundly altered the relations that transform a space into a place. It has changed our way of looking, moving, sitting, it has reset the proximity with other human beings, attention spans, timing...

Anyone who works on the design of space is wondering about the new priorities, the real or virtual measures, the modeling of the new and the reshaping of what already exists.

The design of spaces has also become the design of times, which often densifies experiences, piling them up in a simultaneity of emotions that produce reactions our bodies are not yet completely trained to handle.

This book wants to investigate some of these questions, trying to find possible, plausible answers.

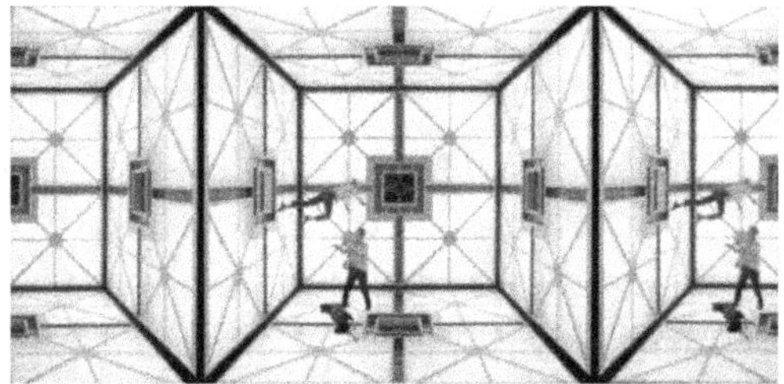

The connection between spaces

is hypertextual, not sequential.

Spatial composition in the digital era

The digital leads to new logics of movement and novel sequences. The composition of spaces obeys other hierarchies that at times are neither functional nor symbolic. Hypertext, to which we have become accustomed thanks to the web, forms other connections that have nothing to do with Cartesian geometric forms.

The connection between spaces is hypertextual, not sequential. Moving in the web is a procedure of connective weaves that have no beginning or end, endowed with a body that has no gravity, but shifts with speed and lightness.

From the web we learn that it is possible to enter from any door, in any moment, and to pass from one world to another following an infinite range of choreographies.

The entrance is not the only access, but simply the main one. Buildings that arise from digital thinking are Piranesian spaces, with a completely subverted principle of orientation.

The web has introduced us to other forms of aggregation, other priorities, values that suggest continuous movement and constant connection as the sole proofs of existence.

In the end, it has opened up spaces that lie elsewhere, those that now contain our data, archives of memories and files, clouds that make our baggage lighter, our movements more agile, allowing us to work anywhere.

The web is not another world
that exists parallel to ours, but an
unforeseen expansion of the real
world we have inhabited for years.

Webbing

The web is not another world that exists parallel to ours, but an unforeseen expansion of the real world we have inhabited for years.

The idea of conquering other planets implied precisely this, but instead of going to Mars we have gone into the network, in any case another infinity. Nevertheless, the web is a space of our life and what happens there, to all effects, has repercussions on our emotional and physical experience.

In *The Game* Alessandro Baricco coins the term "*webbing*" to indicate that way of staying in both worlds – real and virtual – in a seamless way.

In practice, being in both worlds changes the relationship between the body and surrounding space. Perhaps in the future it will mean redesigning that space, also because that is what is relentlessly already happening.

The light in the spaces of webbing is dimmer on average, a sort of twilight that makes the light of the screen more vivid, oriented so as not to create reflections and glare.

Think about how our relationship with space has changed since we have begun to constantly have a device (laptop, smartphone, tablet) before our eyes, or now that we listen to music at high volume on buds pressed into our ears, which simulate the architecture of a surrounding space that seldom coincides with the physical space around us.

The need for movement in spaces is diminished, and sometimes by staying still and extending through the spaces of sensors and headsets we virtually roam spaces without having to walk, without having to be there inside them.

we virtually roam spaces

without having to walk,

without having to be there

inside them.

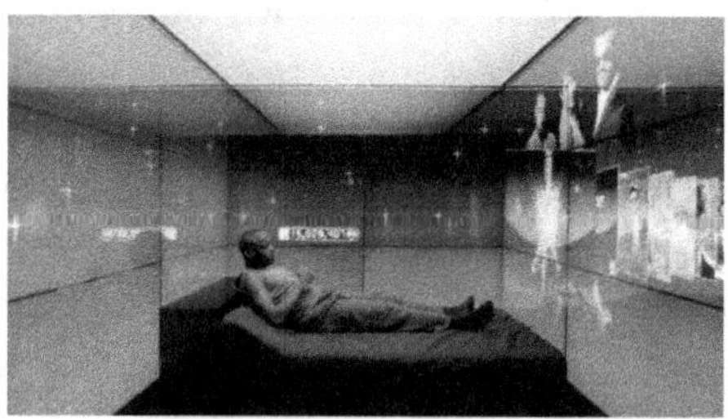

Real spaces are emulating

digital places.

Overstimulation

This time machine is very powerful because it allows us to enter the digital world, but at the same time it allows that world to enter our lives with a force of immanence few people are able to resist. The arrival of e-mail or a message often takes priority over any action in progress, which we interrupt to see the content that has just arrived.

Years ago I read *On the High Wire* by Philippe Petit, who wrote about the necessity when walking a tightrope of paying no attention to anything around you, shutting out any stimuli that might arrive from the context in order to concentrate only on yourself and the destination to the reached: "every thought on the high wire is a fall waiting to happen."

So every time I interrupt what I am doing to react to new e-mail or messages I think about the continual threats to concentration caused by this insistent stimulation. We live in the era of compulsive response, and our bodies are getting used to it.

The spaces in which we exist undergo a barrage of external prompts that distract us, saturating places with stimuli. This is why certain spaces boost the tone of their presence with very noisy sonic backdrops, dazzling lights, bright colors, aggressive odors. Real spaces are emulating digital places.

The new frontier of Spatial Design has to do with the radical interpretation of this idea, taking stimulation towards full immersion. The concept of making experience complete, aided by a "sensorialistic" and not "sensorial" use of our perceptive faculties, leads to the theatrical spectacle typical of theme parks and hotels.

To an increasing extent, it begins with a film, a successful TV series, a cartoon, with their arsenal of characters, settings and stories. Everything else becomes emulation, cosplay, staging, with a ticket to pay at the entrance and a merchandising industry to feed.

design students today know how to use certain tools better than their teachers

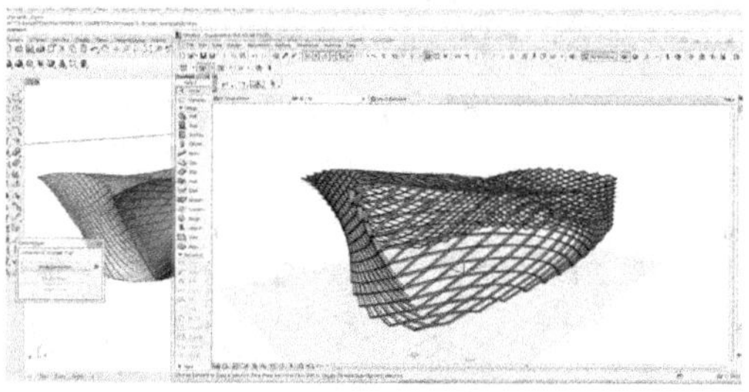

the real question is how to redesign design, not simply to learn how to use the "new drafting machines"

Teaching Spatial Design

It is clear to everyone that for the first time in the history of education, design students today know how to use certain tools better than their teachers. This assertion is true, but also incomplete.

While students know how to use these new and indispensable tools – the programs demanded by the market – much better than their professors (on the average), they do not know what to use them for, what content they should set out to produce, and how to evaluate it.

The most advanced and active universities subject the faculty to constant requirements of technological and linguistic updating, in order to adapt teachings to the modes of learning and interaction of the digital native generations.

The opportunity is compelling because in practice the real question is how to redesign design, not simply to learn how to use the "new drafting machines."

This concept is fundamental in order to understand the meaning of the whole book, and I would like to underline it: we cannot design the way we did before the advent of software, and we cannot proceed simply by learning how to use it. We have to rethink our methods, inventing new ones and new forms of teaching, education and design, because the problem is not a matter of mere translation, but of methodological reformulation.

The new design software not only boosts the possibility of modeling, but also extraordinarily augments the possibilities of design in all its parts, and all the ways of teaching Spatial Design. Just like what happened from the dawn of perspective to the present, every new tool modifies our way of formulating places.

Where should we start today if we want to design a space? What are the contexts of reference, real or digital? How can we conduct research in the age of Pinterest? How can we analyze a place from a video? Are the comments on social networks posted by users important items of information for a designer?

Teaching Spatial Design today means working with students inside a situation full of equipment, from the most basic items like pencils and paper to the most recent, like 3D printers. Every piece of equipment is useful to develop and enhance certain parts and not others. Making a digital model is useful in order to understand forms, but it does not replace the physical model that is used to understand forces, materials, etc.

Part 2

#designer

Scene: A German supermarket recently made a unique gesture of protest. In response to the spreading phenomena of racism against immigrants, the store decided to leave only products made in Germany on its shelves. The outcome, disconcerting for the shoppers and for those like me who saw the images online, was discouraging. Only a handful of products remained on the shelves, namely the ones that were locally produced. The store offered just a few turnips and some potatoes. Otherwise it was empty. Unaccustomed to this unusual "embargo," the customers thought the staff of the supermarket had forgotten to replenish its stock, or that strikes had prevented the arrival of new supplies. But then they noticed the signs scattered throughout the store, which its owners had taken the trouble to prepare. Some of the signs said: "this is how empty a shelf is without foreigners." Others said: "Everything is so boring without variety."

Our students belong to a generation of designers who move in the world, building their expertise through ongoing cultural contaminations. This is their signature, and this is our wager.

Made in...

The students of the Workshop in Spatial Design, which I teach together with colleagues and assistants at the Milan Polytechnic, come from 30 different countries. They left their homes and came to Milan to study design. We communicate in a sort of Anglo-Italo-Hispanic-Franco-Chinese-Korean-Nippon-Indian language, but we all speak the visual/sensory language of design. I teach them to use their senses when they have to define the qualities of space. We do research on books from traditional libraries, printed and online magazines, websites, design libraries on the web... When we find a fascinating example and fall in love with it, we try to take a low-cost flight to visit, explore and understand it, to learn about its qualities with our bodies. In recent years we have understood that the social media gather stories, emotions, the comments that inhabitants and visitors post regarding their experiences inside that space. We discover the life of the place we are studying through their comments. Then we try to look at it, not just in photographs, but also using our smartphones to make small videos, to record movements inside spaces, and to move inside them. Lately we also try to hypothesize how you could design by observing spaces from drones, a bird's-eye view, imagining how we will move in a future that is perhaps not so distant.

Every person designs spaces starting with their own cultural background, opening it to the academic, methodological and design teachings provided by an Italian university, though one of international range. Our students belong to a generation of designers who move in the world, building their expertise through ongoing cultural contaminations. This is their signature, and this is our wager.

I often wonder: will my Chinese, Russian or Inuit students, trained in an Italian university by a team of teachers and professionals who work all over the world, become "Italian" designers at the conclusion of their studies? Will they too be products of "Made in Italy"? Then I reconsider, because the more pertinent question might be: in a couple of years, will it make any sense to ask something like that?

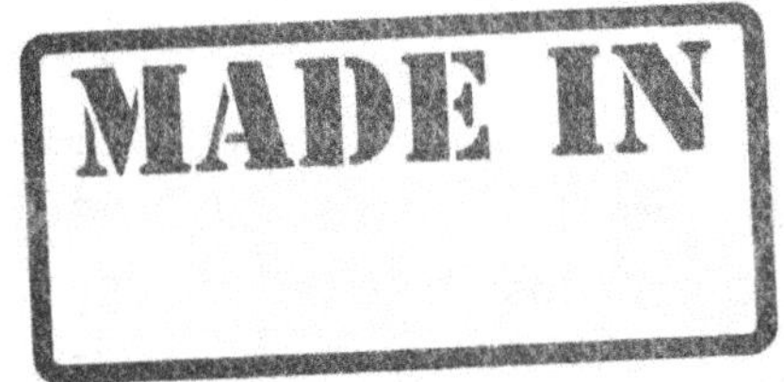

Nothing is more harmful than

becoming your own brand.

Authors and creative commons

I have been teaching for 25 years, Architecture first, and now Spatial Design. I have taught in many universities and many parts of the world. Everywhere, I have noticed an initial reluctance regarding the practice of project teamwork.

Working and studying in groups is the most instructive and important exercise a design school can offer, because design practice involves creativity as a personal problem-solving attitude, not an objective on its own.

In effect, the figure of the author conserves its mythical status even in the most collaborative era in the history of creativity, which is our era: if I say Apple I think of Steve Jobs, if I say Facebook I think of Mark Zuckerberg, if I say Arduino I think of Massimo Banzi.

This is a breaking point: the designer of this millennium can no longer be a solitary author. A project in the digital age is never the result of a virgin birth, but of the efforts of multiple "parents."

In design the figure that signs its name is not a person, but represents a team, a system of people, beyond the ego of the leader of the pack.

In practice, the authorial myth is a double-edged sword: on the one hand it leads the market to fall in love with a designer-brand; on the other, it fastens that figure to a language, demanding and re-demanding only that stylistic formalism that can become suicidal mannerism. Nothing is more harmful than becoming your own brand.

Also in this context, the arrival of the Internet has totally shuffled the deck. On the web you can find anything, every idea has already been posted, someone in some part of the world has already thought about a project like yours. In product design this is very clear, but also in the design of spaces it can be blatantly observable – just open any page on Pinterest.

Design doesn't have to go looking for the umpteenth innovative idea. It has to look for innovative meanings, as Roberto Verganti writes in his latest book *Overcrowded*, seen as "a novel vision that redefines the problems."

The web is the richest and most generous global market of ideas, many of which are also free. Those who design can start there to develop a critical capacity of interpretation, selection and curation.

The web is the richest and most
generous global market of ideas,
many of which are also free. Those
who design can start there to develop
a critical capacity of interpretation,
selection and curation.

Opensource

When the open source concept began to circulate a few years ago, I was asked to review a book by Carlo Ratti and Matthew Claudel titled *Open Source Architecture*, which was a very optimistic dissertation on what might happen in the near future in the field of collaborative Spatial Design.

The question left unanswered by the book, which still bothers me today, is: if I fell ill and needed an operation, would I want to see a highly acclaimed and qualified surgeon by my bedside, or a crowd of global advisors with all kinds of different expertise, ready to take a shot at fixing my problem?

The answer goes without saying. Nevertheless, it would be anachronistic to overlook the possibility that amidst the anonymous advisors there might be experts capable of contributing to solve the problem.

The web is an extraordinary place, but it will not make me into a surgeon, nor will it make someone into a designer who has not studied to be a designer.

The open source dimension permitted by the web is the natural result of the radical efforts of the 1970s to put designers, producers and users around the same table.

The web has taught us that if we are generous in real life we will probably be willing to share things in virtual life as well; that the economics of reputation is just as enticing as that of capital; that barter can be more convincing than finance; that the Internet is a gigantic time bank; that we prefer to be the "investors" from the ground up in a good idea than the paying audience at an ugly spectacle; that any project

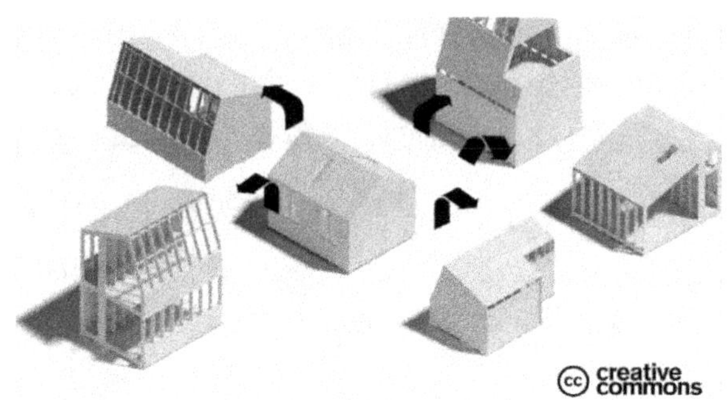

sharing works well when it sticks to
the ethic of the nudist colony: if you
want entry you have to get naked
too, otherwise you are just a voyeur
and you have to stay outside

on the web is a living form capable of being controlled, corrected and continuously updated; that sharing works well when it sticks to the ethic of the nudist colony: if you want entry you have to get naked too, otherwise you are just a voyeur and you have to stay outside.

But then, does the "Linux of dwellings, offices and libraries" exist? Copyleft, Creative Commons, Arduino, MakerBot, Free Beer, RepRap: more than answers, these are redefinitions of the true question, or maybe I'm mistaken. What emerges as extraordinary is the fact that the web is more Pop than Warhol and the replica, the copy, the reiteration, the interpretation add to the project, rather than diminishing it, in keeping with the new formula "the more you copy me the more I exist (at least in the web)."

In architecture the first experiments are still in the testing phase (though some are almost 10 years old): the open source architecture (2006) of Cameron Sinclair, who theorizes disciplined collaboration among designers online (open 24 hours), scattered around the world, capable of responding to and revising needs in real time; WikiHouse, which in 2013 a very young Alastair Parvin presented in a TED talk, which makes a library of designs generated by users available to print, cut and assemble; as well as Brickstarter, Geode, Estate Guru... but we're just at the beginning.

©COPYRIGHT

Ↄ COPYLEFT

success is no longer a matter
of originality, but of speed of
implementation and market availability

Open knowledge_design

Jack Ma, founder of Alibaba, had this to say some time ago about copies of products: "they are often better than the originals, cost less and come from exactly the same factories." While this is not completely honest – there was a clear conflict of interest at work – it is not so far from the truth.

In practice, when western companies began to openly produce things in countries like China, South Korea and Vietnam, etc., then bringing the results back to their own markets to be sold at the same old local prices, they began a process not only of production transfer, but also and above all of know-how transfer. A protectionist viewpoint that sees what has happened as simple copying and patent infringement fails to fully grasp the transmission that has taken place.

The phenomenon of the Chinese *shanzhai* involves products that bypass intellectual property laws. It began about 20 years ago among manufacturers working on different parts of the same product, and as Silvia Lindtner of the University of Michigan writes, it expressed "a culture of sharing of know-how among makers, comparable to the open source phenomenon."

What emerges from this type of culture is that success is no longer a matter of originality, but of speed of implementation and market availability. An understanding of this leads to a new relationship (still partially to be reformulated) between the project and its production, prior to the entry of products on the market.

The birth of the Wiki world, which defines itself as a "website on which users collaboratively modify content and structure directly from the web browser," has the same rules. Arduino also reflects a similar desire to replace the authorial dimension with collaboration, connecting designers, makers (sometimes the designers themselves) and users in a seamless system.

This implies the acceptance in design of Open Knowledge, making it into a system that can be continuously implemented, which will undoubtedly lead to intriguing results.

Internet of Things

Years ago Bruce Sterling wrote: "When you bury fiber-optic under the curbs around the town, then you get internet. When you have towers and smartphones, then you get portable ubiquity. When you break up a smartphone into its separate sensors, switches, and little radios, then you get the internet of things." The Internet of Things arises as the outcome of an initial wager on the part of Bill Gates, who about 40 years ago set out to put a computer on everyone's desk. We can say that he achieved his goal, and this led to another wager: to put a computer inside everything. Everyone has had the experience of looking for a wallet, a book, a set of keys, and wishing it were possible to enter a hypothetical ⌘F (FIND) command and suddenly see the thing blinking for easy location.

In recent years the dialogue between people and spaces, furnishings and objects has intensified, making spaces more efficient and more prone to personalization. But the Internet of Things meets with resistance due to the growing need for security and privacy inside spaces. Just consider the fact that at the moment one of the products setting sales records in the world is the robot vacuum cleaner disk, which cleans your house but also has to map it in order to optimally function. During this detailed survey, it not only takes the measurements of spaces, but also monitors their form and our habits, sending all the information to a central headquarters that "calculates

to put a computer inside everything

your vacuum cleaner
sends the map of your
home and the report on
your habits somewhere

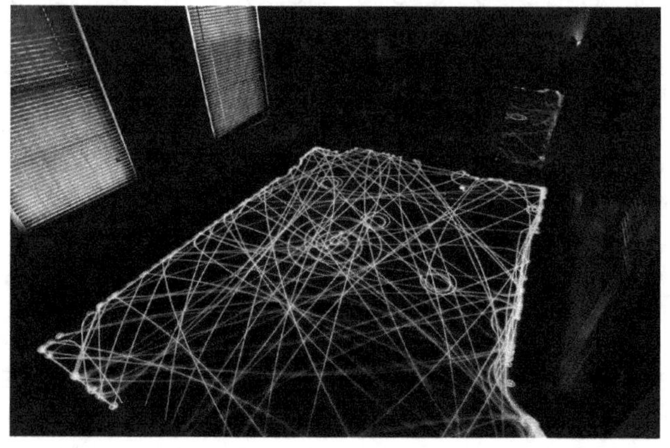

you have already given
your consent somehow,
without noticing it

its route with an algorithm that processes data from sensors on the bumpers and other IR sensors." In substance, your vacuum cleaner sends the map of your home and the report on your habits somewhere.

The American company that has been selling these robots since 2002 claims that it will never sell the information thus gathered, but it has also said that its idea is to "share the data free of charge with the consent of the customer." This means that you have already given your consent somehow, without noticing it. It is an increasingly widespread vision shared by companies and users that an ecosystem of objects and services capable of simplifying our lives, communicating with each other and implementing actions on their own, has to be created inside our homes.

This is an inexorable process, because today the insertion of a microchip in any object during the production phase, in order to connect it to the Internet, has a minimal cost, and because in practice the advantage of being able to turn on the home before we arrive, of allowing the refrigerator to tell the supermarket that we are out of milk, and so on, ensures a sense of care and service that many people are already willing to purchase by giving up their (by now) rather blurry notion of privacy.

Part 2

#place

Scene: I'm in Dubai. Outside it is dreadfully hot and humid, the air unbreathable. I just left a mall that contains the biggest aquarium of the Persian Gulf and Ski Dubai, an unusual ski trail that allows local residents and tourists to ski on artificial snow inside a construction about a hundred meters away from the open desert. In these places everything is air conditioned, and even during the milder months people do not live outdoors, but scurry from one sheltered place to another enclosure. I enter the hotel lobby, humming with AC, at a temperature almost 20 degrees lower than the air outside. The difference forces me to wear a jacket, which I can luckily roll up and stick in my handbag, ideal for travel in these countries.

I go up to the room, and as is often the case in international hotels the AC is running, set to ice cold. But there is a thermostat that lets you adjust by a few degrees, or even turn it off.

That would solve the problem, were it not for the fact that the bed is covered by a very heavy quilt, of the kind found in chilly Nordic countries, utterly unsuitable for the local climate but perfectly in tune with the artificial polar refrigeration produced by the air conditioning.

I wonder what sense there is in this system: cold air and a quilt inside, very hot air and a t-shirt outside. The answer is simple: every hotel chain finds it easier to handle a single set of bedding that works for everyone in all seasons. So

they seal the windows shut, making them impossible to
open, and pump up the AC to a constant chill 365 days a
year, making you sleep under a hot duvet to avoid getting
frostbite.

De-context

The web is making many pipe dreams come true, and one of them is certainly the desire for ubiquity, to be in more than one place at the same time.

I have already extensively explored the vertigo of ubiquity in *Sensi, tempo e architettura* (Time/Sense Based Architecture), a book I wrote a few years ago on the temporal revolution, but what emerges in a piercing way is that this scampering from one place to another leads to syncopated movements never sensed before, and above all profoundly alters the experiential and narrative sequence with which we move around in real spaces.

Movement has completely different rules in digital space. When we change position in real space there is a contextualization of themes, people, things; when we move in the web things are near or far away depending on a different geometries. Mobility without movement is the typical way of roaming the web.

There is no approach along the path to join one point to another, not even a horizon, because the context seems to sink into a very deep spatial fold. I find what I am looking for – a piece of information, an image, a person – according to algorithms that cannot be deciphered, but offer me a selected range from which to choose, not a map with which to get my bearings.

Mobility without movement is the
typical way of roaming the web.

The same thing happens in interpersonal relations: before, if you liked someone, while getting up the courage to introduce yourself you scoured far and wide among their friendships, the places they spent time, their love affairs... Today there are apps that let you meet someone for intimate relations by looking through a catalogue of candidates on hand in the zone where you are presently located. The context is erased, every person is an island. Who is the person in front of you? Who are their friends? What is their job? Their story, plausibly, is a biography of digital images carefully posted to show them in the best possible light. This lack of coordinates is perhaps the vertigo into which we are initiated by the digital natives.

Their story, plausibly, is a
biography of digital images
carefully posted to show them
in the best possible light.

Paradoxes of globalization

Why has globalization so radically altered the form of the world? For a complex series of reasons, some of which I will list here:

- The outermost layer of the earth, the one regarding information, communication, transport... has been crumpled, folded, abandoning physical geography to embrace financial and political geography. The distances between points of the real world are continuously reset by market logics, that make every city more proximate to another city than to its own environs, so Milan is closer to Paris than it is to Cremona. This is why a Milan-Paris airplane ticket costs less than Milan-Cremona, and the time to get from Milan to Cremona is proportionately longer and more expensive (1 hour 54 minutes, costing 30 euros, 101 km) than a trip from Milan to Paris (1 hour 19 minutes, cost 29 euros, 852 km).

the distances between points of the real world are continuously reset by market logics, that make every city more proximate to another city than to its own environs

- A crumpled world produces other proximities and other distances. This is why space takes on different configurations, certain parts sinking, other slipping, others almost vanishing in the deeper folds. We are not accustomed to it, but it is what has been happening for years now, with social, anthropological and design impact still to be determined. In a crumpled world the most proximate place is not necessarily the closest one. Those who are not in the network, those who do not have access, have fallen into a chasm, a valley from which it is impossible for most to climb back out.

- Real and virtual are no longer parallel worlds, but intertwined realities, both capable of triggering emotions.

In a crumpled world the most proximate place is not necessarily the closest one. Those who are not in the network, those who do not have access, have fallen into a chasm, a valley from which it is impossible for most to climb back out.

- The relationship with time is not sequential but experiential, or as the geographer Luc Gwiazdzinski writes, à *la carte*, namely malleable depending on needs, on demand by compulsive consumers.

- The separation between perception and emotion implies a detachment between what I perceive and what I feel, which I sense because it happens in a deferred time.

- The compromised relationship with nature gives up the specificities of a place, also in terms of climate, and tries to reconstruct artificial environments everywhere, completely disconnected from the place.

"Communicating technologies tend to
change the form of the house and the
city, not because we need a special space,
but because the mentalities and everyday
behaviors change."

(Ida Faré)

This sense of freedom, of eternal mobility,
of access to everything and everybody, of an
amplifier of our thoughts, is what seduces us.

Chessboard or chessmen

In the long years of collaboration at the Milan Polytechnic in the courses taught by Ida Faré on the places of living, the theme of the place and its relations with its inhabitants was a central issue, as was the birth of "new species of spaces" between public and private, no longer able to remain inside the walls of the home. There was also the matter of history and anthropology. At the dawn of the digital age Faré wrote: "Communicating technologies tend to change the form of the house and the city, not because we need a special space, but because the mentalities and everyday behaviors change."

The advent of the digital is a major revolution not only for those who design, but also for those who inhabit. The places we live in are hybrids between "ancient" spaces (even the modern is ancient with respect to the digital) and digital modes.

It is not a question of parallel worlds, as was erroneously thought for a long time, but of worlds that coexist in the same spaces. The impact of the new media does not only modify the pieces involved in the game, but the entire chessboard, rewriting the game's rules.

The spaces we inhabit are offspring of Cartesian geometry, with fixed walls and immobile objects, while in the computer, or in the smartphone always within reach, the objects and data move with other weights, light and mobile.

This sense of freedom, of eternal mobility, of access to everything and everybody, of an amplifier of our thoughts, is what seduces us. Digital spaces are so powerful from the viewpoint of messages and emotions that we demand the same performances from both existing spaces and new spaces.

And while on the one hand we enter the web to seek and experience new spaces, in the meantime the same digital windows opened in our homes and offices allow worlds to enter, deferred or simultaneous.

Spaces have become scenarios to post, but they are also outposts from which to act in the network, that allow us to be present where we cannot physically be, permitting us to have breakfast with our kids when we are on the other side of the world, attenuating the sense of separation and distance felt by those who live far away from their loved ones.

I remember the world of the first Erasmus students, without e-mail, lining up at phone booths on holidays in their countries, inserting heaps of coins as if they were playing slot machines.

Today all human beings can live where they want to if they have the connection, they can close it and open it to their habitat, making multiple places and landscapes coexist in the same space.

In practice, perhaps the use we make of the Internet is an effect and not the cause of another new desire to inhabit places.

Today all human beings can live where they want
to if they have the connection, they can close
it and open it to their habitat, making multiple
places and landscapes coexist in the same space.

John Thackara, in *How to Thrive in the Next Economy: Designing Tomorrow's World Today*, writes, "In 1962 [...] Thomas Kuhn introduced the term 'paradigm shift' to describe the ways that scientific worldviews periodically undergo radical change in what appears at the time to be a sudden leap. These 'sudden' paradigm shifts in worldview follow years, sometimes decades, in which scientists have encountered anomalies that don't fit in with the dominant paradigm."

So to walk down the street and come across a person in a heated discussion with someone else who isn't there, a person who doesn't even see us and bumps into us, is the epiphany of a paradigm shift to which we have become accustomed. Likewise, communicating with your children who are in another room in the same house by sending them WhatsApp messages instead of going to call them to dinner defies the paradigm of proximity.

Why do we feel such pleasure when we spend hours lying in bed, chatting with friends instead of going out to meet them somewhere? Or inviting them over? What is the thrill produced by multitasking that allows us to be in multiple places at the same time?

The sensations we have when we move in the web definitely produce an extraordinary thrill we do not want to live without, and maybe it wouldn't make any sense to do so.

Digital places are not the enemies of real places. They are reciprocal extensions destined to intertwine, in some cases to complete each other and – only in the most extreme slippages – to replace each other.

I'd say that over the short term we will still need to go to the bathroom in the morning, to walk, to embrace. Talking with the kids on Skype might make us miss them less, but it cannot replace a big hug.

we want real spaces to be

just as exciting as the web

From inhabitants
to consumers of places

One of the salient issues of Spatial Design is the fact that the experiences we want to obtain from places have changed. The advent of portable media and smart devices has led to a gradual separation between perception and emotion, but also a demand for spaces that offer higher, more exciting, more engaging performance, as if reality should generate the same adrenalin produced by gaming.

To put it bluntly, we want real spaces to be just as exciting as the web.

In this sense we have passed from being inhabitants to being consumers of spaces, seeing space as a product.

The space consumer has a predatory attitude, demanding an increasingly high level of entertainment, interaction and narration capable of channeling one experience or more, incessantly, without an instant of waiting or silence.

Being consumers of places means using values without producing new ones, which can imply cannibalizing resources, failing to build reciprocal relationships with the inhabitants, using the qualities of the space for ourselves, our own interests, and then departing, leaving the place bereft of its resources.

Also for places, there can be responsible or compulsive, useless, bulimic, careless consumption.

Just consider the phenomenon of mass tourism, which transforms cities into consumer goods, expecting them to provide constant performances, ongoing, unnatural at times, out of scale, out of season… exasperating the inhabitants and distorting the very nature of the places.

Globalization has yet to begin

As a result of the availability of information, references, markets and everything else, instead of opening up the world has become standardized. The vertical differences that were previously linked to the taste and culture of a people have been replaced by a large international, horizontal elite that moves extensively but always stays in the same hotels – as long as they have a wi-fi connection – and has a stable economic position, access to new technologies, constantly connected.

The web is a big world, but at the moment it is prevalently filled with things that resemble each other, with a very similar flavor and tone. The libraries from which the market draws images and references are the same. The music heard there has the same identical rhythms and travels on the same frequencies.

The opening of borders has not taken us far away from the world of design into uncharted zones. No, everything has the same taste, everyone wants the same houses, the same accessories, equal all over the world. Because using the same tools makes us feel like part of the same time, the same humankind. A population of designers exists that is equal all over the world and wants to be that way, to distinguish itself, to be recognizable: people who want to drink from the same coffeepots, dress in the same brands, choosing the same magazines as the sole arbiters of "taste."

using the same tools makes us feel like part

of the same time, the same humankind

Globalization has yet to begin, if for the moment it simply means going to produce things in a country with lower labor costs. A global world should become a great opportunity for everyone to go and to touch, listen, come to terms with those who do it better than we do, with those who have different views, those who eat in a different way and listen to other music. We should return from these encounters with globalization laden with an enriched design vocabulary, memorable experiences, odors never smelled before, sounds never heard before. We should experiment with our knowledge in other countries and allow other cultures to experiment in ours. And we should design spaces that include the greatest number of communities, opening any doors to diversity.

In design we are still in a rather kitsch initial phase. It seems as if everyone dreams of living the same life, published in interior decorating magazines. At times we seem to be living in a gigantic duty free store in a generic international airport.

Globalization has yet to begin, if for the moment it simply means going to produce things in a country with lower labor costs.

#community

Scene: I walk into a Chinese store, a workshop where they fix mobile phones and other electric and electronic objects. The young man who runs the place always has a big smile when I enter, but he speaks very limited Italian. He's a highly skilled tech guy, though. Over the last few years he has managed to revive practically all the broken objects I've entrusted to his care. On the counter in the workshop stands a large computer, always on. Not the latest model, since he must have given it hundreds of new lives over time. The computer hasn't been turned off for ten years, connected to someplace in China. Sometimes it is connected to his home, because I can hear domestic noises, and the affectionate tones of an elderly lady who talks as if they were in the same room. Other times it is connected to a big diffused laboratory, some workshop in China or in a Chinatown of another country. He asks for information and someone answers with instructions, recommendations, explaining how to do the repair. As he works, he also gives advice to others, stopping to approach the screen where someone shows him something. He nods and resumes work.

I have never entered his store and found that window closed; that little workshop in Milan has always been connected, like a satellite, to the mothership on the other side of the world.

At lunchtime he doesn't go out. A food delivery service organized by his community brings a meal, absolutely Chinese in its ingredients and preparation.

My friend lives in Milan while still dwelling in China.

Community and communities

The relationship between community and place is another important theme connected with Spatial Design.

Though at the root of many wars still in progress, the idea of a people connected to a nation will soon be extinct, because the most unstoppable process of migration of goods, cultures, information and people ever to take place is now under way.

Various forms of mobility exist, between two extremes: on one side the smart movements of a planetary elite, on the other the biblical exoduses of the remaining majority.

You can live like a Chinaman in Italy, precisely as you could have lived like an Italian in Chicago, without learning English, like my great-grandmother at the start of the 20th century.

Today you can live in another country while eating as if you were in China, buying Chinese ingredients, speaking Chinese, frequenting China while you are in Italy, paying with Chinese currency in stores made for Chinese people, thanks to a Chinese app that prevents any passage of funds on the economic circuits of the host country.

In the digital era certain communities have taken form on the web, bringing together people who did not know each other before, but share various interests. The same people have then met each other in the real world, having a meal (Dîner en Blanc), dancing (Urban Tango), demonstrating (flash mobs), constituting new communities. Other communities have moved towards the web, becoming ghettoized in virtual spaces.

The digital revolution has brought people together in a predictable way, but separated them in unexpected ways. It has scattered certain communities, but it has also created new ones; it has liquefied the boundaries of states, even as some of them build new walls to keep out migrants; it has separated us from our next-door neighbors; it has encouraged sharing among like minds, while failing to guarantee inclusion of diversity.

New names ought to be invented for what is happening. What should we call people who purchase electronic residences in order to become citizens of countries they have never visited? To what state does the German town that presents itself as the first Chinese smart city in Europe belong?

The digital revolution has brought people together in a predictable way, but separated them in

unexpected ways.

it has encouraged sharing among like minds, while

failing to guarantee inclusion of diversity

Glocal schedule

Another paradigm that has multiplied, taking on completely unusual forms, is time. Its transformation is not due exclusively to the digital revolution, though this has made a significant contribution to the process.

The presence of a digital connection in our lives has transformed the screen in front of us into a temporal window, allowing us to access other worlds, other times.

On the web there is a single time, 24/7, a world that is always open and available. The web is a wheel of a gigantic clock that connects the time of the real space in which I exist with that of the real space beyond another screen.

Those who live away from home, like many of my students, may develop a way of existing in two different times zones, that of the classroom with their colleagues, which is the local time, and that of their house, which instead runs on "glocal" time, in synchrony with that of their country of origin. Some of them have told me that on weekends they "eat with their families" who could be in India or China.

The web is a wheel of a gigantic clock that connects the time of the real space in which I exist with that of the real space beyond another screen.

People who work in international networks synchronize their timing with partners and clients living in other time zones. They might have to start work at 5.00 AM to respond to a client in China, and they may continue until late at night, because in the meantime the customers in the USA have had breakfast, and they too must be contacted. It is possible to have a datebook based on different calendars, with weekends from Friday evening to Sunday evening, or others starting Thursday evening and lasting till Saturday night. Some people simply never stop working. In short, a web that is always open and available corresponds to a world that struggles to get into synchrony.

Part 2

#market

Scene: the lobby of a hotel in China. The date is July 15th. I'm waiting for a client who is picking me up to go to a meeting. We're ahead of time, so I have a look around, and then without expecting it I find myself humming the tune of Jingle Bells. I realize that it is the soundtrack of that place – it got into my head – and that I will keep on humming it for the rest of the day.

Something doesn't fit. Because as soon as I hear that ditty a completely different atmosphere takes form in my mind: it's cold out, people are lighting up Christmas decorations hung somewhere, and above all there is a latent sensation that the holidays are coming, which many people (myself included) like very much.

But nothing of all that can be seen in the lobby.

Like the classical symphonies played by corporate answering machines while you're on hold, the simplified, trivialized melodies of ring tones, isolated out of context in time and space.

An everyday paradox that brings irreparable environmental drifts in its wake, but also extraordinary semantic and cultural opportunities. Until the 20th century a form of connection existed between signified and signifier, which in the 21st has become more blurry and transverse. Decontextualization is an important theme for Spatial Design because coherence is no longer required; nor is consistency with the context, or even truth in many cases. The famous *genius loci* is swept away by the concept of the site-specific, fragmented and lacking in temporal sedimentation; truth and authenticity are replaced by plausibility.

in place of mediators the selectors appeared,

who organized the collections on a platform;

or the curators who took the risk of a

personal choice and usually used blogs; the

influencers who presented themselves as

gurus to follow with unshakeable faith

The crisis of mediation

Among the many mechanisms that have faltered with the advent of the web, the crisis of mediation is the most dramatic but also the most stimulating: everyone has access to everything (or almost), on their own.

But if everyone can (virtually) gain access to places, products, producers, research, music, people... if everything is available and within reach, how are we to choose? And above all, how are we to know what is right for us?

Apart from the initial general excitement, the phenomenon immediately seemed disruptive, but with a tendency towards very reckless risk.

Could it be trusted? Who were the people selling or offering services online? Who would guarantee that something was a good deal and not a swindle?

Then came the period of the construction of an ethics of online trade, meaning that if you sold a fraudulent product you could no longer take part in another deal, because the community itself would act as an arbiter and guardian. Soon, however, we realized that even without mediation a form of regulation had to be provided, and in place of mediators the selectors appeared, who organized the collections on a platform; or the curators who took the risk of a personal choice and usually used blogs; the influencers who presented themselves as gurus to follow with unshakeable faith.

In practice the web has brought us closer to what we want, making a myriad of possibilities available; you can just reach out and grab them. In the middle there is no longer any old authority, and the passages have been shortened.

You can get to anyone, anywhere. Celebrities, amazing vacation spots, the incredible adventures of anyone are close at hand if someone has posted them, tweeted them, narrating them day by day as if they were members of our family: just stay connected.

So the job required of the designer is not to generate ideas, but to select and sift through possible or better solutions for the client, with a legitimacy quotient that is decidedly lower than in the past.

so the job required of the designer
is not to generate ideas, but to
select and sift through possible or
better solutions

They called it "reputation"

With the demise of the middlemen our value, that of our works, is provided by reputation. This could function as in the past on the basis of merit, conveyed by the positive feedback of the market, clients, users, inhabitants.

A space is deemed well designed if, for example, people willingly spend time there, or simply because they express positive opinions about it.

When a system of expression of those opinions exists and is public in its operation, these views can become the measure of the value of the project itself.

Thus described, it would appear that the web is the best press office for anyone who has qualities to sell or simply to display.

At first that was the case, namely that whatever had lots of "likes" shot up to the top of the rankings of visibility, while a dearth of likes would make it plunge to the more deeply embedded pages. The web seemed to have put into practice a sort of horizontal assessment of quality, giving us hope for a more transparent world. Over the years, however, a "reputation economy" has developed. It has become a currency, and true or apparent followers or users express views on projects, products and services, orienting market trends simply because they are being paid to do so.

whatever had lots of "likes" shot up to the top

of the rankings of visibility

For websites of the hospitality industry, or even for simple restaurants, this has meant praising or demolishing entire businesses, detouring mass migrations of clients towards one hotel or another, one restaurant or its competitor, exclusively on the basis of reviews that are not necessarily truthful.

In the healthiest of cases, in tune with the practices of the web, reputation is created by the judgments of consumers, clients and users. This assessment is considered more genuine than that of the experts, and less accurate than that of the influencers. The latter – veritable online gurus – are able to shift reservoirs of "likes" to feed the hunger of the market and brands. Without any objectivity of judgment, or any particular expertise (other than that of being super-users), through the social media they steer the powerful, perilous machine of reputation.

To ensure objectivity and to ratify judgments, however, there should be a system of control, evaluation, trust, at least for the selectors. There should be a system in which:

- the selectors are always in good faith and have no conflicts of interest;

- the algorithms of web selection that make the personality or image that has the most "likes" more visible should be considered as indicators of what appeals most, not necessarily of what is best. This means that the "opinion of millions of incompetent people is more reliable than that of an expert, if you are capable of interpreting it";

- what appeals to everyone should also appeal to us, otherwise the effort of searching the web for different offerings becomes a sort of challenge against all the algorithms, which always lead us back to the choice with the most "votes."

Sharing economy

Another paradigm that has been profoundly changed in and because of the digital age is the idea of property. Our relationship with the things and places we possess has changed. The value of property no longer consists exclusively in its possession, but also in its circulation inside a system of sharing. The idea is that if you own an asset it is "mobile"; if you are not using it, someone else can use it and pay you for the service.

The circular character of use – of a car, a home or an office – generates an economy of sharing and produces a quantum leap in the idea of spaces and our way of designing them.

Those who commission the design of a residence no longer do so based on a vision of family life, but in keeping with possible income-generating uses. The fact that entire zones of the home may remain under-utilized, because a child has moved abroad for schooling and returns only for one week each year, or because during vacations the whole house is left empty and "available," seems wasteful today. For economic, cultural and environmental reasons, these assets are "put into circulation."

The value of property no longer consists exclusively in its possession, but also in its circulation inside a system of sharing.

Residential, productive and service-industry spaces have been widely impacted by the possibilities offered by the sharing economy, and this has been taken as input for the design of spaces.

If the space no longer corresponds to an exclusive principle of property, but also to a logic of income, it has to be designed based on different concepts, with different temporal parameters. Above all, its layout has to jibe with practices of extreme versatility and flexibility. Designing a space today also means designing the thousands of possible lives it can narrate.

Part 2

#clientele

Scene: In 2017 Amazon launched a competition to find the city that would host the giant company's second headquarters (HQ2). The stakes included 50,000 jobs, with average salaries of 100,000 dollars per year. But the requirements were very strict: a metropolitan area with over one million inhabitants; a work force from which to be able to recruit specialized experts; a maximum distance of 45 minutes from an international airport; a distance of two kilometers from a highway junction; and access by means of public transportation.

As many as 238 cities took part in the contest, from which two winners were selected (New York and Crystal City, Virginia). Some locations offered significant tax breaks to the company, while others even offered to change their names to incorporate that of the brand (Calgary would become Calmazon or Amalgary). This way of operating with cities – nothing new in the United States – speaks of a world in which physical space will be evaluated more on the basis of digital performance that on that of infrastructural efficiency.

New forms of cities and public spaces will arise from such alliances, similar perhaps to the offices of these companies: playgrounds for adults capable of transforming the workplace into a place for living, eating, sleeping, playing ping-pong, or finding a partner who necessarily has the same kind of life in which to share. Perhaps there will be cities designed to welcome new types of citizens, digital natives immune to the thirst for ownership, oriented towards barter and economies of time. Maybe they will not even use automobiles, letting drones make their deliveries of goods or meals. They will have apps for sharing and cutting down on search time, and they will be attractive, relaxed and rumpled. We cannot say if they will be happy, but they will live longer than we do, aware of what will finally put an end to their lives, thanks to DNA screening.

Special client wanted

In the general magma that constantly moves the pieces in the great game of design, the client too has changed. What remains certain, as it has for several thousand years, is that if the designer does not have a good client, the project will be the outcome of tiresome trials and compromises, which wind up penalizing the work, the designers and the client.

A good client knows how to choose a designer through careful analysis of background and economics, and how to then leave the selected talent free to work without too much interference.

The client makes the difference, at least as much as the designer.

I quote from an article published in 1980 by Pierre-Alain Croset in an excellent issue of *Rassegna*, titled "The Clients of Le Corbusier": "He worked above all for clients that offered him the opportunity to conduct experimentation, or invited him to construct a 'demonstrative architecture.' And if it is possible to assert that his entire artistic life was guided by a true ethic of experimentation, we should not forget that on the margins of his work as an architect we can see the continuous *recerche patiente* for the ideal client." It is clear that without his courageous clients, the genius of Le Corbusier would not have accomplished very much. To back this assertion, we can turn to a book by

The client makes the difference,

at least as much as the designer.

Deyan Sudjic that clearly explains the dynamics of the situation, *The Architecture of Power*, which investigates the inseparable relationship between power and the transformation of spaces.

The world is full of good designers castrated by incompetent or ignorant clients, just as it is full of famous mediocre designers elevated to great heights by brilliant clients and cooperative magazines. And not much can be done in this regard.

The world is full of good designers castrated by incompetent or ignorant clients, just as it is full of famous mediocre designers elevated to great heights by brilliant clients and cooperative magazines.

What clients want

When clients call on a designer, they do it because:

- they don't know what they want but they know what they don't want, so they ask the designer to help them to formulate the questions that are best for them: they want a STRATEGY;

- they have a problem and want an expert to solve it. In this case the designer, like a doctor, has to find a cure, acting as a problem solver in a lucid, technical way: they want a SERVICE;

- they hope to look at the world through the eyes of the designer they have chosen, to create something they are not capable of designing or simply expressing on their own: they want an act of CREATION.

A major misunderstanding exists among all aspiring designers: they think design is only a creative act. One of the first traumas of a young design grad is undoubtedly the discovery that that is not exactly how the world works. Creativity is one part; the rest is professional expertise (aesthetic, technical, commercial, relational and cultural in nature). This very simple, almost banal concept is one of the fundamental stumbling blocks of a great number of designers on our planet.

they want a STRATEGY

they want a SERVICE

they want an act of CREATION

The market demands interpretation, response, formulation or concatenation of meanings in storytelling form. Creativity is the added plus, of inestimable value, the signature through which each designer, faced with the same problem, finds his or her own optimal solution.

Faced by the digital revolution in progress, we have to abandon any form of nostalgia and try to enact the most radical genetic mutation that has ever happened in Spatial Design since its origins: that of giving up on the role of the author and altering its romantic paradigms.

Creativity is the added plus, of inestimable value, the signature through which each designer, faced with the same problem, finds his or her own optimal solution.

Fear of innovation

One of the myths of the 20th century is that the world of design always has to seek innovation, and that success is linked to the achievement of the most innovative solution possible.

This myth reached peaks of fanaticism in product design, in the days of the grand visions of Steve Jobs, capable of dragging even mere mortals into worlds never seen before.

Those visions belonged to an era I would define as the "race to the ultimate invention," not always mainstream. Just consider the smartphone: 2017 was the year in which sales dropped for the first time. In that sector the market is mature, saturated, and the costs have risen in proportion to those of luxury cars. In the future we will be happy to buy earlier models, because the hunger for innovation has already been vigorously sated.

Not only the market but also designers think that projects must always contain an indispensable novelty quotient. We behave like fashion designers in pursuit of a trend to follow, some unusual taste to which to respond. But that is not how things work.

In the life of a designer is becomes immediately clear that not all clients are equal, and that some clients have no need, and/or intention, and/or courage, and/or necessity to set forth on adventures in unexplored realms.

Instead, they ask us to design something that resembles something else, simply interpreting it for their context.

The rendering paradox

A society that is so highly evolved in terms of imagery uses sight to communicate and looks to design, as a result, for plausible and realistic models of spaces, in order to narrate their character as if they already existed.

The abstraction typical of the sketch is banned – especially in interior design – and replaced by renderings/videos that are as polished as they are premature, demanded practically on a daily basis by the client right from the start of the project, to "see what it will be like" and to decide if it fits the bill or not.

This approach is comparable to that of a customer in a restaurant who insists on sampling the pasta just a few minutes after it has been set to boiling in its pot, and then demands constant tastes of its progress, running the risk of finishing the meal before it hits the plate. Such a practice would undoubtedly lead to complaints about ruined repasts.

This anxiety of visualization of the space in its final qualities is one of the biggest problems faced by designers, who have to respond to the demands of a client who approaches the design of a space as if he were examining items to be purchased online, which are already available.

Renderings are forced to follow a reversed model of the design process: they start from the image and then try to extrapolate the places (in the best of cases). Today it seems old-fashioned to try to stand up to the desire to see in advance, and renderings are an effective way of producing a plausible narrative of spaces.

At this point, right or wrong, the client wants to see the project immediately, prior to deciding on assigning a commission. Clients don't want to make the effort of translating the ideas of the designer into their context, so they ask to see the project even before it has been designed or properly considered.

The paradox is the production of a rendering or video that is as realistic as possible, but based on completely arbitrary choices, from the outset, until client approval has been received. The designer remains the victim, because henceforth any reconsidered feature, any variation, has to again be explained and justified.

The websites of design firms are packed with renderings that are so realistic it is hard to distinguish them from photos of finished works, making it impossible for an untrained eye to tell the difference between a design studio and a "render farm."

renderings/videos are as polished as they
are premature, demanded practically on a
daily basis by the client right from the start
of the project, to "see what it will be like"

they often become banal malls of online shopping for aspirations, without understanding that selecting an image of a space is different from imagining a space and then living inside it

Accustomed to the increasingly sophisticated production of images, Spatial Design often falls victim to their charm, sometimes neglecting the participation of the other senses in the design of beauty, and reducing the creation of spaces to a mere photogenic exercise.

Photogenic exercises

Just as we designers access online image libraries, so the market peruses them, browsing, clarifying what is needed and what is wanted. The online platforms that publish projects could be meeting places between designers and clients. Instead, they often become banal malls of online shopping for aspirations, without understanding that selecting an image of a space is different from imagining a space and then living inside it.

Ceci n'est pas une pipe was the title of a work by Magritte in 1926, which then became a serial exercise on the relationship between representation, description and reality. And this is precisely the most frequent expectation – to be able to understand spaces through images.

Copying images does not mean designing places, an activity that instead involves paying attention to light (natural and artificial); to sounds, present and generated; to odors, temperature, the physical touch of surfaces, the visual touch of finishes, the chromatic narrative, humidity in the air, the presence of other human beings...

Accustomed to the increasingly sophisticated production of images, Spatial Design often falls victim to their charm, sometimes neglecting the participation of the other senses in the design of beauty, and reducing the creation of spaces to a mere photogenic exercise.

Instagrammability

Until a few years ago, going to see a beautiful place also implied taking a few snapshots away with you. Today, in that same ritual, the photographer is expected to be the protagonist of the shot, and the place becomes simply a backdrop. The age of selfies is nothing new, but this particular way of relating to places as sets raises interesting issues regarding Spatial Design.

Certain museums, at this point, have become "selfie factories" where the works serve only to bear witness to the effective presence of the visitor in that specific site. The visit per se is less important than the shot to be posted more or less immediately. The Victoria & Albert Museum has even introduced the hashtag "#myvam" to optimize this habit of its visitors. The institution uses their photos for the advertising and promotion of the museum itself.

Instagram and the social media networks influence our projects, because their "social life" has an impact on their success, at least on a par with the impact of their design quality. On the market, companies require designers to supply an increasingly tangible quotient of "instagrammability." Selfie-points act as placemarks destined to become social icons for brands, deployed as publicity.

The Rove Hotel in Dubai, run by the company Stride Treglowe, has decided to put hashtags everywhere, to encourage the hotel's target to be active on the social networks right from the entrance: #this is where I am.

The same logic has gone into the design of the Nike store in Shanghai, where the branding agency Rosie Lee has built a throne of shoes, and the Harrods department store by Farshid Moussavi, who reaches the point of announcing that creating instagrammable moments "is now part of architectural briefs, especially in the world of hospitality."

Anish Kapoor's *Cloud Gate* in Chicago is one of the world's most often posted selfie points. Places are being created with the sole function of acting as iconic settings.

The explicit request of the businessman who commissioned the Vessel by Thomas Heatherwick in New York, for the sum of 200 million dollars, was to have "an Eiffel Tower for tourists." The project, just like that tower, is a maze of staircases in a Piranesian and parametric system leading to nowhere, except to the experience of climbing up, taking a selfie and posting it.

Design for spaces to post on social networks leads to new harmonies, new rules, new canons of post-Vitruvian proportioning: as in the squared spaces of Instagram. The firm Vale Architects recently published an Instagram Design Guide, to help designers to make their interiors more "instagrammable," which is what the market wants.

Places are being created with the sole

function of acting as iconic settings.

Part 3

#senses

The man of 2598 (Bruno Munari)

A scholar of the future (Jean-Pierre Esposito Tung, born in 2598) sets out to reconstruct the type of human being that lived in the period from 1925 to 1971, based on the quality and variety of the seats, armchairs and sofas manufactured in that era. It would appear, especially from the diligence with which these items were approached and designed, as if they were more important than all other issues, that the typical human being living in that time period must have had an enormous backside, swollen by the lack of transpiration caused by the plastic used to build or cover chairs and sofas... So beautiful chairs are designed that make lots of noise when moved, beautiful armchairs that weigh a ton or an ounce, spaces in which air cannot circulate. In the homes created by design odors are not taken into account, as if the body were lacking a nose, so unpleasant scents waft freely through the rooms, the smell of fried fish winds up in the closet, while the aroma of mothballs invades the kitchen. The aspect of noise is utterly ignored: toilets are heard flushing in the parlor, sounds of all kinds echo in all places; not to mention the design of gorgeous, luxurious restaurants where the noise is deafening. If the designer is the coordinator of all the aspects of a project, it is his responsibility if mankind conserves its senses, or if some of them get atrophied. As we know, while functioning develops organs, lack of activity leads to atrophy. Many animals that live in environments different from that of their origin lose the use of certain organs, such as animals that lose the sense of sight by living for a long time in dark caves. Man too can lose the use of certain organs if they are not stimulated. We ought to remember that the individual is equipped with all the senses and that there also is the pleasure of touching agreeable materials, that wood is more pleasant to the touch than metal, that certain materials allow the body to breathe while others do not, that noises can be irritating even when we are used to them, that odors are not always appealing, that the high-pitched screech of certain strollers can ruin the hearing of newborn babes... If, as it appears, functioning develops the organ, then non-functioning will make it atrophy. So in the future will there be men without ears? Or without noses? Or with a back and a bottom deformed by lack of transpiration? Will this be the human being of the future? We hope not. So let's remember that when we design something that also makes good tactile sense, people will notice and will start to use the sense of touch again, which is one of the most neglected senses. If we also take the other senses into account, people will gradually get used to it and will discover that there are many sensory receptors through which to know about the world in which we live. Children know that very well, and their first knowledge of the world is sensorial, global...

During digital interaction the
body is often lacking in spatial
and sensorial awareness of the
surrounding space.

Sensefulness

Digital space is clean, aseptic, without friction, weight, consistency... it is an audio-visual world in which to live, getting beyond the physical limits of the gravitational world.

We are so accustomed to the speed of the web that the idea of standing in line, of waiting, is no longer acceptable. The thrill of the network creates dependency. My digital body is composed of the information about me that exists in the web, without criteria of truthfulness or coherence... and of my body.

When we take part digitally in the world, the body takes on different postures, seated, lying down, or in motion with one hand holding the digital prosthesis that is the smartphone.

During digital interaction the body is often lacking in spatial and sensorial awareness of the surrounding space. No friction, no cold, no weight, no smell. The center of the story is inside the web, though the emotional front is on and inside the body.

It makes little difference if the space is dark, noisy, brightly lit, because the body is emotionally excited more by what happens across the screen than by what happens on this side of it. On the side of reality there is a beached whale, on the virtual side a nimble gazelle that runs in a world that makes every representation theatrical.

The cause-effect relationship between our movements is exponential: a few small gestures correspond to infinite gestures.

Even when we are living in the web, our body is in a physical space, the body is there and the emotions – also those stimulated by the web – have repercussions on it.

The statement might seem quite banal, but instead it is crucial, because although the new technologies continue to be oculocentric, the experience of places seeks an indispensable emotional completion in the other senses.

If you ask normal people (not designers) to describe a place, the story will go beyond the gaze: they will talk about silence, crowding, warmth and coolness, perceptible odors, comfort and so on, in a series of descriptions that report on their experience of that place: an alchemy between objective qualities and personal sensations that constitutes not only their experience, but also the memory that have of that experience.

The central issue is no longer that of attaching identities to things, but that of designing what lies inside them, in their DNA. Identity is not a label that can be added after the fact. Identities are lived, not worn.

Starting over from the experience of places can help to give spaces back to ordinary people. They are assigned the role of actors, not spectators, of the space they inhabit day by day and can/must modify.

On the side of reality there is a beached

whale, on the virtual side a nimble gazelle

that runs in a world that makes every

representation theatrical.

So the "digital revolution" that has altered spaces, breaking down their geometries and modifying their tools of measurement, gives way to a "temporal revolution." This has led to the design of forms of time rather than forms of space, ushering in the era of the sharing economy and raising new questions that are still open to debate. Now the moment comes for a "sensorial revolution," in which the senses become fundamental intermediaries in the integration of real and virtual, and the body lays claim to its role as a terminal of emotions that come from reality and the realm of the virtual at the same time.

now the moment comes for a "sensorial revolution,"

in which the senses become fundamental

intermediaries in the integration of real and virtual

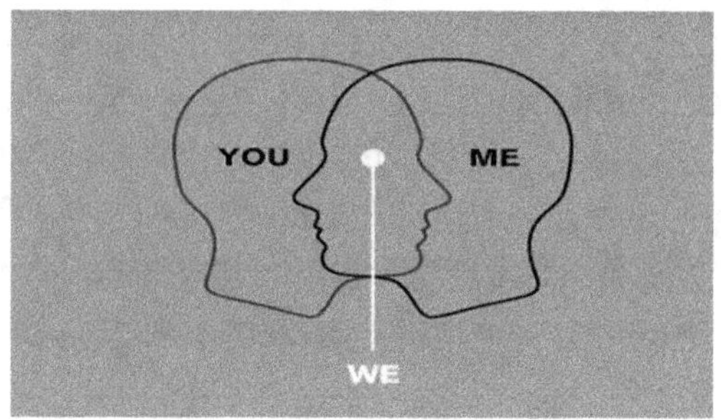

In front of a screen, in the
interaction with machines,
empathic communication gives way
to efficient communication, and we
lose to ability to fully comprehend
our counterparts.

Sense_based design

Some months ago I received a briefing for a Spatial Design project in a foreign country. For the first time in my career, the briefing included a very interesting chart that contained a table of requirements indicated for the individual spaces. The document expressly addressed the design of the experience of places, and systematically listed the needs using four categories: human, sensory, communication, physical.

So – I thought – this was a sign that something was being done to recover a necessity, the lost relationship with places, through the body, the senses, human and physical experience.

The very detailed brief listed the need to have spaces filled with natural light, to have the possibility of putting the body into different positions across the day, experimenting with various postures. The document called for acoustic sensitivity, not only to soften noise but also to transform various areas into sonic environments in which to converse without being overwhelmed by unwanted music. It requested surfaces that would encourage contact, that were not too reflective or glossy, and could play with light in a subtle shaping of the design. It urged that the air not be too "conditioned," and that certain odors marking the identity of the place be allowed to circulate, perhaps allowing users to open some of the windows. Temporal orientation was recommended, helping people to know what time it was in any moment, without having to look at a clock. At night the building was supposed to go on living, with another life, perhaps, granting the designer the honor of envisioning more than one.

I had been waiting for years for this to happen again, for the emphasis to shift away from the sole technological dimension or the communicative qualities of spaces. Deep inside, I had sensed that the body was going to rebel and would leap back onto the stage where it had been the protagonist for thousands of years. It is a different body, in a revised space, but the need to reclaim the physical dimension has returned to the fore. The body is still the location of our emotions; whether they come from the room next door or the web, it is always our body that reacts.

Through the senses we also reconstruct empathy, that human and animal capacity to feel at one with others, to identify with their sensations. In front of a screen, in the interaction with machines, empathic communication gives way to efficient communication, and we lose to ability to fully comprehend our counterparts.

"Our unconscious is becoming a refined master of avoidance of expectation, comprehension and dedicated attention, i.e. the triad of empathic communication," digital strategist Giulio Xhaet wrote in *Il Sole 24 Ore* on 16 October 2018.

Only by designing with the mediation of the senses can the imagination be guided towards other perceptive dimensions and, as a result, those of emotions and empathy.

As never before, the designer's kit required for this new expedition is a sensorial toolbox, ready to aid and interpret the signals of this umpteenth revolution, and above all to encourage the rise of a series of languages and practices already in progress, which lead us toward greater sharing and awareness of the spaces in which we live.

I would like to call this design practice sensefulness, to achieve an immersive experience of places without the need for too many allegories. Likewise, I would like to use the term sense-based design to indicate products, services, brands that foster conscious experiences through the body and its apparatus.

The response to visual overexcitement is a completely sensorial design, in which the senses do not intervene as bogus decoration at the end of the design process, but are instead the constituent tools of the DNA of design and its spaces.

I would like to call this design
practice sensefulness, to achieve
an immersive experience of
places without the need for too
many allegories.

a Ida Faré

Sensefulness

Nuovi paradigmi per le Spatial Design

Anna Barbara

Premessa

Questo libro vuole esplorare cosa sta accadendo nel mondo per via della rivoluzione digitale e le ricadute nel design degli spazi. Per chi, come me, ricerca e insegna in università, ma anche progetta in giro per il mondo, ci sono molte questioni che iniziano a non tornare più. Ho pensato quindi di fermarmi e provare quantomeno a mettere in ordine quelle questioni che si aggirano tra le pagine degli appunti, che prendo durante le revisioni con i miei studenti, mentre ascolto cosa dicono i miei giovani collaboratori, mentre incontro i committenti, mentre scorrono le notizie di un mondo sempre più compresso e spesso più iniquo.

E' un periodo complicato quello che stiamo vivendo. Alcuni diritti ci sembravano acquisiti, per sempre, e invece lasciamo ai giovani in eredità l'urgenza di difenderli con i denti. Questo periodo duro e affascinante impone una condivisione degli scenari. Chi pensa di arroccarsi aldiquà di un qualche muro vivrà assediato, come chi si rifugia nel *web* pensando di poter vivere lì resterà alienato.

Fuori da questo mondo che corre, ci sono quelli rimasti indietro, i più lenti, quelli che hanno saltato un giro su questa giostra vertiginosa, e sono stati eiettati fuori dal mondo del lavoro e non riescono più a rientrare. Ci sono gli anziani che non capiscono che lingua parliamo, gente in coda alle mense dei poveri e altri in palestra a smaltire, ci sono quelli che fan resistenza, ci sono gli attivisti o i populisti, ci sono i poveri, profughi, naufraghi, disperati, in mezzo a un mare, in marcia verso un confine, sul marciapiede di casa mia.

Abbiamo tutti in mano un cellulare, chi per *chattare*, per *postare*, per lavorare o cercare qualcosa; chi per chiedere aiuto ai soccorritori, per segnalare le coordinate e farsi salvare, per avvisare la famiglia da sopravvissuti senza altro passaporto.

Non ho mai creduto per un solo istante, che il mondo di cui scrivo, sia l'unico mondo esistente.

Parte 1 _ il design degli spazi

Ho pensato a come risuonano ogni tanto le lezioni che tengo ai miei studenti, che provengono da tutto il mondo, cui vorrei insegnare ad essere designers e non solo a farlo.

Vorrei aiutarli a capire come essere progettisti di un mondo in trasformazione di cui, alle volte neanche io e i miei colleghi, siamo in grado di offrire interpretazioni convincenti.

Con il medesimo rispetto e responsabilità, ho pensato ai miei committenti che sembrano aver subìto una mutazione genetica da quando ho iniziato a progettare, con cui sento la necessità di riformulare una serie di paradigmi e valori che ci riguardano entrambi.

Cos'è accaduto agli spazi con l'avvento dell'era digitale riguarda anche il design. Tutte le università del pianeta, tutti gli studi di design e i luoghi di riflessione sul progetto, se lo stanno chiedendo.

E' come se le trasformazioni dei "luoghi" siano avvenute più velocemente del pensiero che le avrebbe dovute progettare e ci troviamo a vivere in "spazi", spesso progettati con le logiche dell'abitare del secolo scorso.

Eppure la rivoluzione digitale ha mutato profondamente le relazioni che trasformano uno spazio in un luogo. Ha modificato il nostro modo di guardare, di muoverci, di stare seduti, ha rimisurato la prossimità con gli altri umani presenti, l'attenzione, i tempi...

Chiunque si occupi di design degli spazi si sta interrogando sulle nuove priorità, sulle misure reali e virtuali, sulla modellazione del nuovo e rimodulazione dell'esistente.

Il design degli spazi è diventato anche design dei tempi, che spesso densifica le esperienze, le accavalla in una compresenza di emozioni che producono reazioni a cui il nostro corpo non è ancora completamente allenato.

Questo libro si vuole interrogare su alcune di queste domande e provare a trovare delle eventuali risposte sensate.

Comporre spazi nell'era digitale

Il digitale promuove nuove logiche di movimento e insolite sequenze. Gli spazi si compongono con altre gerarchie che non sono talvolta né funzionali, né simboliche. L'ipertestualità, cui il web ci ha rieducato, disegna altri collegamenti, che non hanno a che fare con le forme geometriche cartesiane.

La connessione tra gli spazi è ipertestuale, non è sequenziale. Muoversi nel web è un procedere in maglie di connessione che non hanno inizio fine, dotati di un corpo che non ha la gravità, spostandosi con velocità e leggerezza.

Dal web impariamo che si può entrare e uscire da qualsiasi porta, in ogni momento e passare da un mondo all'altro secondo infinite coreografie.

L'ingresso non è l'unico accesso, ma solo il principale. Gli edifici che nascono dal pensiero digitale sono spazi piranesiani, con un principio di orientamento completamente sovvertito.

Il web ci ha iniziato ad altre forme di aggregazione, ad altre priorità, a valori che pongono il continuo movimento e la perenne connessione come uniche prove dell'esistenza.

E infine ci ha aperto a spazi altrove, quelli che oggi ospitano i nostri dati, archivi di memorie e di *files*, *clouds* che rendono più leggere le nostre valigie, più agili i nostri spostamenti, che ci consentono di lavorare ovunque.

Webbing

Il web non è un altro mondo che vive parallelo al nostro, ma è un'espansione imprevista di quel mondo reale che abbiamo abitato per anni.

L'idea di conquistare altri pianeti era proprio questa, ma anziché andare su Marte, siamo andati dentro la rete, comunque un altro infinito. Tuttavia la rete è uno spazio della nostra vita e quello che vi accade dentro ha, a tutti gli effetti, una ricaduta sulla nostra esperienza emotiva e fisica.

Alessandro Baricco in *The Game*, conia un termine che è "*webbing*" per indicare quel modo dello stare in entrambi i mondi –reale e virtuale- senza soluzione di continuità.

Nei fatti stare in entrambi modifica il rapporto tra corpo e spazio circostante. Forse in futuro significherà riprogettarlo, anche perché inesorabilmente questo già sta accadendo.

La quantità di luce negli spazi del *webbing* è mediamente più tenue, una sorta di penombra che rende più vivida la luce dello schermo e la direzione è rivolta a evitare fenomeni di abbagliamento.

Pensiamo quanto sia cambiata la relazione con lo spazio da quando abbiamo un media (computer portatile, *smartphone*, o *tablet*) di fronte o quando ascoltiamo la musica ad alto volume nelle orecchie, una musica ad altissima fedeltà che ridisegna l'architettura dello spazio circostante che non sempre coincide con quello nel quale ci troviamo.

Diminuisce la necessità di spostamento negli spazi e talvolta stando fermi, e disseminando gli spazi di sensori e visori, andiamo virtualmente in giro per gli spazi senza necessità di camminare, senza necessità di esserci.

Sovrastimolazioni

Questa macchina del tempo è potentissima, perché ci consente di entrare nel mondo digitale, ma a sua volta di fare entrare quel mondo nelle nostre vite, con un imperativo d'immanenza cui in pochi resistono. L'arrivo di una mail o di un messaggio diventa spesso prioritario a qualsiasi azione in corso, che interrompiamo per leggere il contenuto di quanto è arrivato.

Anni fa mi capitò di leggere il *Trattato del funambolismo* di Philippe Petit, che raccontava la necessità quando si cammina su una corda di annullare l'attenzione su qualsiasi cosa sia intorno, di spegnere tutte le sollecitazioni che arrivano dal contesto per concentrarsi unicamente su se stessi e la meta da raggiungere: "ogni pensiero sul filo è una caduta in agguato".

Ecco, ogni volta che interrompo quello che sto facendo per reagire a una nuova mail o messaggio, penso a continui attentati alla concentrazione che l'insistente sollecitazione comporta. Siamo nell'era della risposta compulsiva, cui il nostro corpo si sta assuefacendo.

Gli spazi quindi che viviamo, subiscono una tempesta di stimoli esterni che ci distraggono, che saturano i luoghi di stimolazioni. Per questa ragione alcuni luoghi alzano il tono della loro presenza con sottofondi sonori molto rumorosi, oppure con luci abbaglianti o colori sgargianti e odori aggressivi. Gli spazi reali emulano i luoghi digitali.

La nuova frontiera dello *Spatial Design* è legata alla radicalizzazione di questa idea portando la stimolazione fino all'immersività. Il concetto è rendere completa l'esperienza, coadiuvati da un uso "sensorialistico" e non sensoriale dei sensi, che porta ad una teatralizzazione tipica dei parchi e degli hotel a tema.

Sempre più spesso si parte da un film, da una serie televisiva di successo, da un'animazione, con le loro dotazioni di personaggi, ambienti e storie. Tutto il resto diventa emulazione, *cosplaying*, messa in scena, con un biglietto da pagare in entrata e un'industria di *merchandising* da rifocillare.

Insegnare Spatial Design

E' evidente a tutti che, per la prima volta nella storia della formazione, oggi gli studenti di design sanno usare alcuni strumenti meglio dei loro insegnanti. L'affermazione è tanto vera quanto incompleta.

Gli studenti sanno usare questi nuovi irrinunciabili strumenti, i programmi che il mercato richiede, molto meglio dei loro professori (mediamente), tuttavia non sanno per cosa usarli, per produrre quali contenuti, come valutare questi.

Le università più evolute e attive sottopongono il corpo docente a richieste continue di aggiornamenti, tecnologici, linguistici, in modo da adeguare gli insegnamenti ai modi di apprendimento e interazione delle generazioni native digitali alle quali insegnano.

L'occasione è avvincente perché nei fatti la vera questione è riprogettare la progettazione, non banalmente imparare a usare "i nuovi tecnigrafi".

Questo concetto è fondamentale per capire il senso di tutto il libro e lo ribadisco: non si può progettare come facevamo prima dell'avvento dei *software* solo imparando a usarli, dobbiamo ripensare a nuovi metodi, a nuove forme di insegnamento, di educazione, di design, perché il problema non è di semplice traduzione, ma di riformulazione metodologica.

I nuovi *software* di progettazione, non potenziano solo le possibilità di modellazione, ma aumentano straordinariamente anche quelle del progetto in tutte le sue parti e dei modi di insegnare il design degli spazi. Com'è avvenuto dall'avvento della prospettiva ad oggi, ogni nuovo strumento modifica il nostro modo di formulare i luoghi.

Da dove partire oggi per progettare uno spazio? Quali sono i contesti di riferimento, quelli reali o quelli digitali? Come si conduce una ricerca all'era di Pinterest? Come si analizza un luogo da un video? I commenti sui social pubblicati dagli *users* sono informazioni importanti per un progettista?

Oggi insegnare a progettare uno spazio significa lavorare con gli studenti dentro un laboratorio pieno di attrezzature, dalle più primarie come carta e matita, alle più recenti come le stampanti 3D. Ogni attrezzatura è buona a sviluppare e potenziare delle parti e non altre. Fare un modello digitale serve a capire le forme, ma non sostituisce un modello fisico che serve a capire forze, materiali, ecc.

Parte 2 _ #designer

Scena: In questi giorni un supermercato tedesco ha compiuto un gesto di protesta unico. In risposta ai dilaganti fenomeni di razzismo contro gli immigrati, ha deciso di lasciare sugli scaffali esclusivamente prodotti made in Germany. Il risultato, sconcertante tanto per i clienti, tanto per chi come me ha visto le immagini in rete, è stato desolante. Solo una manciata di prodotti erano rimasti sugli scaffali, precisamente quelli che vengono prodotti in Germania. Nel supermercato non era rimasta che qualche rapa e delle patate. Per il resto, il vuoto. Disabituati a questo singolare "embargo" i clienti hanno pensato che il supermercato si fosse dimenticato di rifornire i propri espositori, oppure che ci fossero stati degli scioperi che avessero impedito i rifornimenti, ma poi i clienti hanno notato i singolari cartelli disseminati per tutto il negozio che la proprietà si era presa la briga di scrivere. Alcuni recitavano "Senza stranieri uno scaffale è vuoto", altri ancora: "Tutto così noioso senza varietà".

Made in...

Gli studenti del Laboratorio di Spatial Design, che tengo con colleghi e collaboratori al Politecnico di Milano, appartengono a 30 nazionalità diverse. Hanno lasciato i loro paesi e sono arrivati a Milano a studiare design. Comunichiamo attraverso una sorta di anglo-italo-ispanico-franco-cino-coreano-nippo-indiano, tutti però parliamo le lingue visuali/sensoriali del design. Insegno a loro a usare i sensi quando devono definire le qualità dello spazio. Facciamo ricerca sui libri delle biblioteche tradizionali, sulle riviste stampate e online, sui siti, nelle *libraries* di design nel web... Quando troviamo un esempio affascinante e ce ne innamoriamo, proviamo a prendere un volo *low cost* per andare a visitarlo, esplorarlo, capirlo, imparare con il corpo le sue qualità. Negli ultimi anni abbiamo capito che i *social media* raccolgono i racconti, le emozioni, i commenti che gli abitanti, i visitatori "postano" in merito alla loro esperienza dentro quello spazio. Scopriamo la vita del luogo che stiamo studiando attraverso i loro commenti. Poi proviamo a guardarlo, non solo dalle fotografie, ma anche a riprenderlo attraverso gli *smartphone* che consentono di fare piccoli video, di registrare gli spostamenti negli spazi, di muoversi dentro. Ultimamente proviamo anche a ipotizzare come si possa progettare osservando gli spazi dai droni, a volo d'uccello, come ci muoveremo in un futuro forse non troppo distante.

Ognuno progetta gli spazi a partire dal proprio bagaglio culturale, aprendolo agli insegnamenti accademici e metodologici e progettuali appresi in una università italiana, sebbene internazionale. I nostri studenti appartengono a una generazione di designer che si muovono nel mondo, costruiscono le loro competenze attraverso continue contaminazioni culturali. Questa è la loro cifra, questa la nostra scommessa.

Mi domando spesso, ma i miei studenti cinesi, russi, eschimesi... che studiano presso un'università italiana, con un team di docenti e professionisti che lavorano ovunque nel mondo, alla fine del loro percorso di studi saranno designer italiani? Saranno un prodotto del *made in Italy*? Poi ci penso e forse la domanda più corretta sarebbe: che senso avrà tra un paio di anni porsi una domanda del genere?

Autori e *creative commons*

Insegno da 25 anni progettazione, architettura prima e design degli spazi ora. Ho insegnato in molte università e in molte parti del mondo. E ovunque mi è capitato di cogliere un'iniziale insofferenza rispetto alla pratica di progettare insieme ad altri.

Lavorare e studiare in gruppo è l'esercizio più pedagogico e importante in una scuola di design, perché l'esercizio di progettazione pone la creatività come attitudine personale alla risoluzione dei problemi e non come obiettivo.

In effetti la figura dell'autore rimane mitizzata anche nell'era più collaborativa della storia della creatività come la nostra: se dico *Apple* penso a Steve Jobs; se dico *facebook* penso a Mark Zuckerberg, se dico Arduino penso a Massimo Banzi.

Questo è un punto di frattura: il designer di questo millennio non può più essere un autore solitario. Un progetto nell'era digitale non è mai figlio di una vergine, ma di un lavoro di più "genitori".

Nella progettazione chi firma non è una persona, ma rappresenta un team, un sistema di persone, aldilà dell'ego del capo branco.

Il mito dell'autorialità nei fatti è un'arma a doppio taglio, da un lato porta il mercato a innamorarsi di un designer-brand, dall'altro lo inchioda a quel linguaggio chiedendogli e richiedendogli reiteratamente solo quel formalismo stilistico che può diventare manierismo suicida. Niente è più nefasto del brand di sé stessi.

Anche in quest'ambito, l'avvento di internet ha fatto saltare tutti i pezzi della scacchiera. Su internet si trova tutto, qualsiasi idea è già pubblicata, qualcuno da qualche parte del mondo ha pensato a un progetto come il vostro e l'ha già pubblicato. Nel design di prodotto è molto evidente, ma anche nel design degli spazi è presente in maniera plateale, basta aprire qualsiasi pagina di *Pinterest*.

Il design non deve ricercare l'ennesima idea innovativa, ma significati innovativi, come scrive Roberto Verganti nel suo ultimo libro *Overcrowded*. L'innovazione di significato è intesa come "una visione originale che ridefinisce il valore assegnato al problema".

La rete è il più ricco e generoso mercato globale delle idee, molte di esse sono anche gratis, chi progetta può partire da lì dotandosi di una capacità critica, d'interpretazione, di selezione, di curatela.

Opensource

Quando qualche anno fa iniziò a circolare il concetto di *open source*, mi fu chiesto di recensire il libro di Carlo Ratti e Matthew Claudel dal titolo *Architettura Open Source*, che era una dissertazione assai ottimista di quanto sarebbe accaduto da lì a breve sul tema della collaborazione nel design degli spazi.

La domanda alla quale il libro non rispondeva e che ancora oggi mi assilla è: ma se fossi ammalata e dovessi farmi operare, vorrei al capezzale del mio letto operatorio un chirurgo di riconosciuto curriculum o una pletora di consiglieri globali dalle più variegate esperienze, pronti a intervenire sui miei malanni?

Il chirurgo indubbiamente. Tuttavia sarebbe anacronistico trascurare la possibilità che tra gli anonimi consiglieri ci possano essere altrettanti esperti in grado di contribuire alla risoluzione del problema.

La rete è un luogo straordinario, ma non trasforma me in un chirurgo né qualcuno in un progettista se non ha studiato per diventare tale.

La dimensione *open source* che il web consente è lo sviluppo naturale degli sforzi radicali degli anni '70 per mettere intorno al medesimo tavolo designers, produttori e *users*.

La rete ci ha già insegnato che se siamo generosi nella vita reale, probabilmente saremo propensi a condividere anche in quella virtuale; che l'economia della reputazione ci entusiasma quanto quella del capitale; che il baratto ci convince più della finanza; che internet è una gigantesca banca del tempo; che ci piace di più essere "investitori" dal basso di una buona idea, che pubblico pagante di un brutto spettacolo; che qualsiasi progetto in rete è una forma vivente in grado di essere controllato, corretto e aggiornato in continuazione; che la condivisione funziona bene quando aderisce all'etica del villaggio di nudisti: se vuoi entrare ti svesti anche tu, altrimenti sei solo un guardone e stai fuori.

Ma allora esiste "il Linux delle abitazioni, degli uffici e delle biblioteche?". Il *copyleft*; le *creative commons*, *Arduino*, la *maker-bot*, la *Free Beer*, la *RepRap* più che risposte sono ridefinizioni della domanda vera, ma forse mi sbaglio. Quello che emerge di straordinario è che la rete è più *pop* di Warhol e la replica, la copia, la reiterazione, la declinazione aggiungono, anziché togliere, al progetto, secondo la nuova formula "più mi copi, più esisto (almeno nella rete)".

In architettura i primi esperimenti sono anch'essi in rodaggio (anche se alcuni hanno quasi 10 anni): l'OAN (*Opensource Architecture* 2006) di Cameron Sinclair, che teorizza collaborazioni disciplinate tra progettisti online (*open 24 hours*), disseminati in giro per il mondo in grado di rispondere/revisionare in tempo reale alle esigenze; *Wikihouse*, che nel 2013 un giovanissimo Alastair Parvin presentò a un TED, in cui si mette a disposizione una *library* di progetti generati dagli utenti, da stampare, tagliare e assemblare; poi ancora *Brickstarter*, *Geode*, *Estate Guru*, ma siamo solo all'inizio.

Open knowledge_design

Jack Ma, fondatore di Alibaba, dichiarava qualche tempo fa, rispetto alle copie dei prodotti: "sono migliori degli originali, costano meno e provengono esattamente dalle stesse fabbriche". Quanto affermava non è completamente vero -dato l'evidente conflitto d'interessi- ma è molto vicino alla verità.

Nei fatti quando in maniera palese le aziende occidentali sono andate a produrre in paesi come la Cina, la Corea del Sud, il Vietnam, etc. per poi riportarsi indietro dei prodotti da vendere nei propri mercati a prezzi occidentali, hanno iniziato un processo non solo di trasferimento di produzione, ma soprattutto di passaggio di *know-how*. Una visione protezionistica che pensa che quanto sia avvenuto è un processo di semplice copia e appropriazione indebita dei brevetti, non coglie la trasmissione avvenuta.

Il fenomeno dello *shanzhai* cinese indica i prodotti che aggirano le leggi sulla proprietà intellettuale. E' nato circa 20 anni fa tra produttori che lavoravano su parti diverse di un medesimo prodotto e, come scrive Silvia Lindtner della Università del Michigan, esprimeva "una cultura della condivisione del sapere tra i fabbricanti, paragonabile al fenomeno *open source*".

Quello che emerge da quel genere di cultura è che il successo non è più solo questione di originalità, ma di velocità di realizzazione e d'immissione sul mercato. Comprendere questo comporta un nuovo rapporto (da riscrivere ancora in parte) tra il progetto e la sua produzione, prima dell'ingresso dei prodotti sul mercato.

La nascita del mondo *Wiki*, che per autodefinizione è "un website dove gli *users* in maniera collaborativa, modificano contenuti e struttura direttamente dal *browser*" ha la stessa regola. Anche Arduino aveva il medesimo desiderio di sostituire l'autorialità con la collaborazione, connettendo designers, *makers* (talvolta di stessi designers) e users senza soluzione di continuità.

Si tratta dell'apertura del design all'*Open Knowledge* che rende il design, un sistema implementabile in continuazione di cui ne vedremo ancora delle belle.

Internet delle cose

Anni fa Bruce Sterling scriveva: "Quando installi la fibra ottica sotto i marciapiedi di una città, ottieni internet. Quando hai grattacieli e *smartphone*, ottieni l'ubiquità portatile. Quando scomponi uno *smartphone* in sensori, interruttori e radioline, ottieni l'internet delle cose".

Internet delle cose nasce come conseguenza di una prima scommessa, di Bill Gates, che circa quarant'anni fa si era messo in testa di mettere sulla scrivania di ogni persona un computer. Possiamo dire che ci sia riuscito, e questo ha dato speranze a un'altra scommessa: mettere il computer dentro ogni cosa.

A ognuno di noi è capitato almeno una volta di cercare il portafogli, un libro, un mazzo di chiavi e avere il desiderio di premere un ipotetico comando ⌘F (FIND) e improvvisamente vederlo lampeggiare e ritrovarlo.

Nel corso degli ultimi anni, il dialogo tra noi e gli spazi, gli arredi, gli oggetti si è intensificato per rendere gli spazi più efficienti, più personalizzabili. Tuttavia l'internet delle cose trova delle resistenze per via della crescente necessità di sicurezza e di *privacy* dentro gli spazi. Si pensi che al momento, uno dei prodotti più venduti al mondo, l'aspirapolvere-robot a disco, non solo pulisce casa nostra, ma per farlo al massimo delle sue prestazioni, ha bisogno anche di mapparla. Durante questa operazione di rilievo dettagliato, oltre alle dimensioni dello spazio, monitora anche la sua forma, le nostre abitudini e inoltra tutte queste informazioni ad un quartier generale centrale che "calcola il suo itinerario tramite un algoritmo che elabora i dati dei sensori posti sul paraurti e su altri sensori IR". In sostanza il vostro aspirapolvere manda da qualche parte la mappa di casa vostra e il racconto delle vostre abitudini.

La società americana che vende dal 2002 questi robot, ha dichiarato che non venderà mai i dati raccolti, di contro però ha affermato che la sua idea è di "condividerli gratuitamente con il consenso dei clienti". Il che significa che da qualche parte avete già acconsentito senza prestarci attenzione. E' una visione sempre più condivisa, da aziende e *users*, che si debba creare all'interno delle case un ecosistema di oggetti e servizi in grado di semplificarci la vita e di parlarsi e sbrigarsela tra di loro.

E' un processo inesorabile, perché oggi inserire in qualsiasi oggetto in fase di produzione un *microchip*, che lo connetta a internet, ha un costo minimo e perché in pratica il vantaggio di accendere casa prima del nostro arrivo, consentire al frigorifero di segnalare al supermercato che ci serve il latte, etc., garantisce un senso di cura e di servizio che in molti già sono disposti a barattare con la loro ormai labile idea di *privacy*.

Parte 2 _ #luogo

Scena: Sono a Dubai, fuori fa un caldo infernale, umido, aria irrespirabile. Ho appena lasciato un mall che ospita l'acquario più grande del Golfo Persico e lo Ski Dubai, un'insolita pista da sci, che consente ai residenti emiratini e ai turisti di andare a sciare su una neve artificiale, racchiusa in una costruzione a un centinaio di metri dal deserto. Da queste parti, tutto è climatizzato, e anche durante le stagioni più miti la gente non vive all'aperto, ma passa da un luogo confinato ad un altro luogo chiuso. Entro nella lobby dell'albergo, climatizzata, con una temperatura di quasi 20 gradi in meno rispetto all'esterno. Questa escursione mi obbliga a indossare una giacca che fortunatamente tengo arrotolata nella borsa, quando vado in giro in questi Paesi.

Salgo in camera, e come spesso accade, negli alberghi internazionali, l'aria condizionata è accesa a temperatura glaciale. C'è un termostato e si può regolare di alcuni gradi o addirittura spegnere.

Il problema sembrerebbe risolto, se non fosse che il letto è coperto da un pesantissimo piumone, di quelli che si trovano nei paesi freddi del Nord, inadeguato rispetto alle temperature locali, ma perfettamente coordinato con quelle artificiali e polari che il condizionatore dell'aria impone.

Mi chiedo il senso di questo sistema: aria fredda da condizionatore e piumone all'interno, aria caldissima e maglietta a maniche corte all'esterno. La risposta è semplice: ogni catena di alberghi ha maggiore convenienza nella gestione di un unico set biancheria, che vada bene per tutti e per tutte le stagioni. Così si serrano le finestre, evitando che qualcuno le possa aprire, si accende a manetta il condizionatore fino a una temperatura fredda costante, durante tutte le stagioni dell'anno, e si dorme sotto un caldissimo piumone che ci protegge dal gelo.

De-contesto

Il web sta realizzando molti sogni che avevamo nel cassetto e sicuramente uno di questi, è l'essere ubiqui, essere in più luoghi simultaneamente.

Ho già parlato ampiamente in *Sensi, tempo e architettura*, un mio libro scritto anni fa sulla rivoluzione temporale, della vertigine dell'ubiquità, ma quello che emerge in maniera lancinante è che questo zampettare qui e lì, porta a movimenti sincopati mai avvertiti prima e soprattutto modifica profondamente la sequenza esperienziale e narrativa con cui ci spostiamo negli spazi reali.

Muoversi nello spazio digitale ha regole completamente differenti. Quando ci spostiamo nello spazio reale, c'è una contestualizzazione dei temi, delle persone, delle cose; quando ci spostiamo nella rete le cose sono vicine o lontane secondo una geometria differente. La mobilità senza il movimento è lo spostarsi tipico della rete.

Non c'è avvicinamento nel percorso di unione tra un punto e un altro, non c'è neppure orizzonte, perché il contesto, sembra essere sprofondato in una piega spaziale molto profonda. Arrivo a trovare quello che sto cercando –una informazione, un'immagine, una persona- secondo algoritmi indecifrabili, che mi offrono una selezione tra cui scegliere, ma non una mappa con cui orientarmi.

Avviene così anche per le relazioni tra le persone: se ti piaceva qualcuno, prima di trovare il coraggio di presentarti, perlustravi in lungo e in largo le sue amicizie, i posti che frequentava, le sue storie d'amore... Oggi esistono delle *app* che consentono di incontrare qualcuno con cui avere intimità, sfogliando un catalogo di candidati presenti nella zona dove ti trovi in quel momento. Il contesto è annullato, ognuno è un'isola. Chi è la persona che hai di fronte? Chi sono i suoi amici? Che lavoro svolge? La sua eventuale storia è una biografia d'immagini digitali accuratamente pubblicate con la migliore rappresentazione di sé. L'assenza di coordinate è forse la vertigine a cui i nativi digitali ci stanno iniziando.

Paradossi della globalizzazione

Perché la globalizzazione ha modificato così radicalmente la forma del mondo? Per una serie complessa di ragioni di cui elenco rapidamente alcune:

- il *layer* più esterno della terra, quello che riguarda l'informazione, la comunicazione, i trasporti... si è accartocciato, si è piegato, ha abbandonato la geografia fisica per abbracciare quella finanziaria e politica. Le distanze tra i punti del mondo reale, sono ridisegnate in continuazione da logiche di mercato, che pongono ogni *city* più prossima a un'altra *city*, *anziché* alla propria periferia. Milano è più prossima a Parigi che non a Cremona. Per questa ragione un biglietto aereo Milano-Parigi costa meno che un Milano-Cremona e il tempo per andare da Milano a Cremona è proporzionalmente molto più lungo e costoso (1 ora e 54 minuti costo 30 euro, 101 km) che per andare da Milano a Parigi (1 ora e 19 minuti, costo 29 euro, 852 km);

- un mondo accartocciato definisce altre prossimità e altre distanze. Per questa ragione lo spazio assume differenti configurazioni, alcune parti precipitano altre slittano, altre quasi scompaiono abissandosi nelle pieghe più profonde. Non siamo abituati ma è quello che accade ormai da anni con impatti sociali, antropologici, progettuali ancora tutti da definire. In un mondo accartocciato il luogo più prossimo non è necessariamente il più vicino. Chi non è nella rete, chi non ha accesso, è precipitato in una gola, in una valle impossibile, ai più, da risalire;

- reale e virtuale non sono più mondi paralleli, ma realtà intrecciate in grado di suscitare emozioni entrambi;

- il rapporto con il tempo non è sequenziale, ma esperienziale o, come scrive il geografo Luc Gwiazdzinski, à *la carte*, ossia malleabile a seconda dei bisogni a richiesta di consumatori compulsivi;

- la separazione tra percezione ed emozione, comporta un distaccamento tra ciò che percepisco e ciò che provo, che sento, perché avviene in un tempo differito;

- il rapporto compromesso con la natura rinuncia alle specificità anche climatiche di un luogo e prova a ricostruire ovunque ambienti artificiali, completamente scollegati dal luogo.

La scacchiera o gli scacchi

Nei lunghi anni di collaborazione al Politecnico di Milano, nei corsi di Ida Faré incentrati sui luoghi dell'abitare, il tema del luogo e delle relazioni con i suoi abitanti era un tema centrale, così come la nascita di "nuove specie di spazi" che stavano tra il pubblico e il privato e non riuscivano più a rimanere composti dentro le mura domestiche. C'era il tema della storia e dell'antropologia. All'alba dell'era digitale scriveva Faré: "Le tecnologie comunicanti tendono a mutare la forma della casa e della città, non perché abbiano bisogno di uno spazio speciale, ma perché mutano la *mentalità* e i *comportamenti* quotidiani".

L'avvento del digitale è una grande rivoluzione non solo per chi progetta, ma anche per chi abita. I luoghi che viviamo sono ibridazioni tra spazi "antichi" (anche il moderno lo è rispetto al digitale) e modi digitali.

Non è una questione di mondi paralleli, come erroneamente si è pensato per molto tempo, ma di mondi che convivono nei medesimi spazi. L'impatto dei nuovi media non modifica solo i pezzi coinvolti nel gioco, ma coinvolge l'intera scacchiera e riscrive tutte le regole del gioco.

Gli spazi che abitiamo sono figli di geometrie cartesiane, hanno pareti fisse e oggetti immobili, mentre nel computer, o nello stesso *smartphone* a portata di mano, gli oggetti, le informazioni si muovono con altri pesi, leggeri e mobili.

Questo senso di libertà, di eterna mobilità, di accesso a tutto e a tutti, di amplificatore dei nostri pensieri è quello che ci seduce. Gli spazi digitali sono così potenti dal punto di vista dei messaggi e delle emozioni, che richiediamo agli spazi esistenti e a quelli nuovi le medesime performance.

E mentre da un lato entriamo nella rete a cercare e sperimentare nuovi spazi, nel frattempo le medesime finestre digitali aperte nelle nostre case e nei nostri uffici, fanno entrare mondi, differiti o simultanei.

Gli spazi sono diventati scenari da "postare", ma anche sono postazioni da cui agire nella rete, che ci permettono di presenziare dove non potremmo esserci fisicamente, che ci consentono di fare colazione con i nostri figli quando siamo dall'altra parte del mondo, che stemperano molto il senso di nostalgia e distanza per tutti coloro che vivono lontani dai propri affetti.

Me lo ricordo il mondo dei primi studenti Erasmus, senza email, in coda alle cabine telefoniche nei giorni di festa del proprio paese... a infilare gettoni come nelle slot machine.

Oggi qualsiasi umano può vivere dove vuole se ha la connessione, può chiuderla e aprirla sul proprio abitare, può far coesistere più luoghi e paesaggi nel medesimo spazio.

Nei fatti forse l'uso che facciamo di internet è un effetto e non la causa di un altro nuovo desiderio di abitare i luoghi.

John Thackara, in *Progettare oggi il mondo di domani*, scrive "il cambiamento di paradigma secondo Thomas Kuhn, nel 1962, avviene con "improvvisi" cambiamenti in cui per anni gli scienziati riscontrano anomalie che non rientrano nel paradigma dominante".

Ecco, camminare per strada e incrociare una persona che animatamente dialoga con qualcun altro che non è presente e che neanche ci vede e ci sbatte contro, è l'epifania di un paradigma modificato cui ci siamo ormai abituati. Idem, dialogare con i propri figli che sono in un'altra stanza della medesima abitazione, mandando loro messaggi *whatsapp*, anziché andare di persona a chiamarli mette in crisi il paradigma della prossimità.

Perché abbiamo un piacere enorme a trascorrere delle ore sdraiati sul letto a *chattare* con gli amici anziché uscire fuori e incontrarli? Invitarli a casa? Quale ebbrezza ci produce il multitasking che ci consente di stare in più luoghi simultaneamente?

Le sensazioni che abbiamo muovendoci nella rete ci danno sicuramente delle vertigini straordinarie cui non vogliamo più rinunciare e forse non avrebbe neanche senso farlo.

I luoghi digitali non sono nemici di quelli reali. Sono estensioni reciproche destinate a integrarsi, in alcuni casi a completarsi e -solo nelle derive più estreme- a sostituirsi.

Direi che nell'orizzonte più breve avremo ancora bisogno di andare in bagno la mattina, di camminare, di abbracciarci. Parlare con i figli distanti via skype riduce la nostalgia, ma non sostituisce i loro abbracci vigorosi.

Da abitanti a consumatori di luoghi

Una delle questioni salienti che coinvolge il design degli spazi è che sono mutate le esperienze che noi cerchiamo di ottenere dai luoghi. L'avvento dei media trasportabili, dei *device smart*, ci ha portato a una graduale separazione tra percezione ed emozione, ma anche a una richiesta rivolta agli spazi di essere performativi, eccitanti, coinvolgenti come se la realtà dovesse stimolare la medesima adrenalina che il *gaming* produce.

Messa in maniera provocatoria, vogliamo che gli spazi reali siano eccitanti come il *web*.

In tal senso siamo passati dall'essere abitanti ad essere consumatori di spazi e a considerare lo spazio come un prodotto.

Il consumatore di spazio ha un atteggiamento predatorio, richiede ad esso un quoziente di intrattenimento sempre molto alto, la necessità di interazione, di narrazione in grado di veicolare una o più esperienze, senza smettere mai, senza un istante di attesa o di silenzio.

Consumatori di luoghi significa usarne i valori senza produrne di nuovi, può voler dire cannibalizzare le sue risorse, non costruire relazioni reciproche con gli abitanti, usare le qualità dello spazio per sé, per i propri interessi e andarsene lasciando il luogo depredato delle proprie risorse.

Anche per i luoghi esiste sia il consumo consapevole sia quello compulsivo, inutile, bulimico e privo di cura.

Pensiamo al fenomeno del turismo di massa, che trasforma le città in generi di consumo, che chiede a queste performance costanti, continue, alle volte innaturali, fuori scala, fuori stagione… e esaspera gli abitanti e la natura stessa dei luoghi.

La globalizzazione deve ancora iniziare

In conseguenza a questa disponibilità d'informazioni, referenze, mercati, di tutto, è accaduto che invece di aprirsi, il mondo si sia omologato. Alle differenze verticali, che prima erano legate al gusto e alla cultura di un popolo, si è sostituita una grande *elite* internazionale, orizzontale, che si muove moltissimo, ma soggiorna negli stessi alberghi -purché ci sia un *wi.fi*- che ha una posizione economica stabile, che ha accesso alle nuove tecnologie... e che è costantemente connessa.

Il web è un grande mondo, ma al momento prevalentemente di cose che si assomigliano, che hanno un sapore molto simile. Le *libraries*, da cui il mercato attinge immagini e referenze, sono le medesime. La musica che è ascoltata suona a ritmi identici e transita sulle stesse frequenze.

Non è accaduto che l'apertura dei confini ci portasse lontano in zone inesplorate dal mondo del design. No, tutto sa dello stesso sapore, tutti vogliono le stesse case, gli stessi accessori, uguali in tutto il mondo. Perché usare gli stessi strumenti fa sentire parte del medesimo tempo, della medesima umanità. Esiste un popolo di designer che è uguale in tutto il mondo e vuole essere tale per distinguersi e riconoscersi: vuole bere dalle stesse caffetterie, vestire gli stessi brand ed eleggere gli stessi *magazine* come gli unici detentori del "gusto".

La globalizzazione deve ancora iniziare, se questa significa semplicemente andare a produrre in un paese dove si pagano salari più bassi. Un mondo globale deve diventare una grande opportunità per tutti di andare e toccare, ascoltare, confrontarsi con quelli più bravi di noi, con quelli che la pensano diversamente, con quelli che mangiano in un'altra maniera e ascoltano dell'altra musica. Dobbiamo da questi incontri con la globalizzazione, tornare indietro con un vocabolario progettuale arricchito, con esperienze memorabili, con odori mai annusati e suoni mai ascoltati. Dobbiamo, sperimentare le nostre conoscenze in altri paesi e lasciare che altre culture sperimentino nel nostro. E dobbiamo progettare spazi che includano il maggior numero di comunità e di *community*, aprendo qualsiasi porta alla diversità.

Nel design siamo ancora in una prima fase un pò *kitsch*. Sembra che tutti sognino di vivere le medesime vite, pubblicate nei *magazine* di interni. Sembra talvolta di vivere in un gigantesco *duty free* di qualsiasi aeroporto internazionale.

Parte 2 _ #comunità

Scena: Entro nel negozio cinese, è un laboratorio di riparazioni di cellulari e altri oggetti elettrici ed elettronici. Il giovane signore che lo gestisce ha sempre un sorriso generoso nei miei confronti, ma parla poco e male l'italiano. E' un abilissimo tecnico però. A memoria, negli ultimi anni è riuscito a riparare qualsiasi oggetto io gli abbia consegnato affinché lo resuscitasse. Tiene sul bancone del laboratorio, sempre acceso, un grande computer. Non è l'ultima versione, ma sicuramente lui deve avergli già restituito centinaia di vite oltre la propria. Il computer è acceso, sempre, da dieci anni, ed è collegato da qualche parte della Cina. Talvolta è connesso con casa sua, perché sento rumore di domesticità e toni affettuosi di una vecchia signora che parla come se fossero nella stessa camera. Altre volte è collegato ad un grande laboratorio diffuso, qualche laboratorio in Cina o in qualche china-town di altri paesi. Lui chiede informazioni e qualcuno risponde istruzioni, consigli, spiega come compiere la riparazione. Mentre lavora, anche lui impartisce consigli ad altri, si ferma, si avvicina allo schermo, qualcuno gli mostra qualcosa, annuisce e riprende.

Non sono mai entrata nel suo negozio che quella finestra non fosse aperta, che quel minuto laboratorio a Milano non fosse collegato, come uno spazio satellite, alla sua casa-madre dall'altra parte del mondo.

All'ora di pranzo non esce da lì, una rete di food delivery organizzato dalla sua comunità gli consegna il cibo, assolutamente cinese negli ingredienti e nella preparazione.

Il mio amico vive a Milano ma "abita" in Cina.

Comunità e community

Il rapporto tra comunità e luoghi è un altro importante tema collegato al design degli spazi.

Pur essendo all'origine di molte guerre ancora in corso, l'idea di popolo collegato a una nazione si estinguerà molto presto, perché è in corso il più inarrestabile processo di migrazione di merci, di culture, d'informazioni e persone.

Esistono varie forme di mobilità tra due estremi: da un lato gli spostamenti *smart*, di una *elite* del pianeta e dall'altro gli esodi biblici della restante maggioranza.

Puoi abitare in Italia da cinese, esattamente come potevi vivere da italiano a Chicago senza imparare l'inglese come la mia bisnonna all'inizio del XX secolo.

Oggi puoi stare in un paese, mangiando come se fossi in Cina, comprando gli ingredienti cinesi, parlando cinese, frequentando la Cina in Italia, comprando con moneta cinese nei negozi gestiti da cinesi, pagando con un'*app* cinese che non transita moneta neanche per sbaglio nei circuiti economici del paese ospitante.

Nell'era digitale alcune *community* sono nate in rete, tra gente che non si conosceva, ma che condivideva passioni. Poi le medesime si sono fissate appuntamenti nel mondo reale, si sono incontrate, hanno pranzato (*Dinner in white*), danzato (*Urban Tango*), manifestato (*flash-mob*) insieme, costituendo comunità nuove. Altre invece si sono trasferite nella rete, ghettizzate in spazi virtuali.

La rivoluzione digitale ci ha aggregato in maniera prevedibile, ma separato in modi inaspettati. Ha disgregato alcune comunità, ma ne ha creato di nuove; ha reso liquidi i confini degli Stati mentre alcuni costruiscono nuovi muri per fermare i migranti; ci ha separato dal nostro vicino di casa; ha promosso la condivisione tra simili, ma non ha garantito l'inclusione tra diversi.

Bisogna inventare nuovi nomi per definire quanto accade. Come chiamare i cittadini che comprano residenze elettroniche per diventare cittadini di paesi dove non sono mai stati? A quale Stato appartiene la cittadina tedesca che si candida ad essere la prima *smart city* cinese in Europa?

Orario glocale

Un altro paradigma che si è moltiplicato assumendo forme completamente insolite è il tempo. La sua trasformazione non è dovuta esclusivamente alla rivoluzione digitale, ma indubbiamente questa ha contribuito significativamente a metterla in atto.

La presenza nelle nostre vite di una connessione digitale ha trasformato lo schermo di fronte a noi in una finestra temporale, che ci consente di accedere ad altri mondi e altri tempi.

Nella rete c'è un tempo unico, un giorno continuo, open24/7, c'è un mondo sempre aperto e disponibile. Il web è una ruota di un gigantesco orologio che raccorda il tempo dello spazio reale in cui mi trovo, con quello dello spazio reale aldilà di un altro schermo.

A chi vive fuori sede, come a molti dei miei studenti, capita ad esempio di abitare due fusi orari differenti, quello in aula, con i propri colleghi che è il tempo locale e poi quello a casa, dove vige invece il tempo "glocalizzato", sincronizzato con quello del loro paese di origine. Alcuni di loro mi hanno confessato che nei weekend "mangiano con la propria famiglia" che magari sta in India o in Cina.

Del resto anche chi lavora in un network internazionale, sincronizza i suoi tempi con i partner e i clienti che vivono in altri fusi orari. Può capitare di iniziare a lavorare alle 5 del mattino per rispondere a un cliente in Cina, per poi continuare fino a tarda notte perché intanto si sono svegliati i clienti negli USA e bisogna rispondere anche a loro. Oppure può accadere di avere un'agenda che segue diversi calendari, con weekend dal venerdì sera alla domenica sera, altri che iniziano il giovedì sera fino alla sera del sabato e altri ancora che non smettono mai. Insomma a un web sempre aperto e disponibile corrisponde un mondo che con fatica prova a sincronizzarsi.

Parte 2 _ #mercato

SCENA: lobby di un albergo - Cina. E' il 15 luglio. Aspetto un cliente che venga a prendermi per andare a una riunione. Siamo in anticipo e quindi mi guardo intorno, poi, mentre meno me lo aspetto inizio a canticchiare Jingle Bells. Mi accorgo che è il tappeto sonoro di quel luogo -mi è entrato in testa- e continuerò a cantarlo per tutto il giorno.

Non mi torna. Perché appena sento quel motivo, nella mia mente si dipinge un'atmosfera completamente diversa: fa freddo, si accendono le decorazioni natalizie appese, e soprattutto c'è una sensazione latente di festa alle porte, che a molti (me inclusa) piace parecchio.

Invece di tutto questo, nella lobby non c'è nulla. Come quelle sinfonie classiche, che intrattengono le nostre attese nelle segreterie telefoniche di molte aziende, che suonano sequenze musicali semplificate, banalizzate a suonerie, decontestualizzate nel tempo e nello spazio.

E' un paradosso quotidiano che porta con sé derive ambientali insanabili, ma anche straordinario opportunità semantiche e culturali.

Fino al XX secolo, esisteva una forma di connessione tra significato e significante, che nel corso del XXI è diventata oggi più labile e trasversale. La decontestualizzazione è un tema importante del design degli spazi perché non è più richiesta la coerenza, né la fedeltà al contesto, né spesso la verità. Il fatidico genius loci è spazzato via dal concetto di site specific frammentato e senza sedimento temporale; la verità e l'autenticità sostituite dalla verosimiglianza.

La crisi della mediazione

Tra i tanti meccanismi che sono saltati con l'avvento del web, la crisi della mediazione è il più drammatico, ma il più stimolante: tutti accedono a tutto (o quasi) volendo, da soli.

Ma se tutti possono garantirsi di accedere a luoghi (virtualmente), a prodotti, a produttori, a ricerche, a musica, a persone... se tutto è disponibile e raggiungibile, allora come faccio a scegliere? E soprattutto come faccio a sapere cosa è buono per me?

A parte una prima eccitazione generale, il fenomeno sin da subito è sembrato dirompente, ma con una tendenza al rischio assai azzardata.

C'era da fidarsi? Chi erano le persone che vendevano o offrivano servizi online? Chi mi garantiva che l'affare non era invece una truffa?

Venne quindi il periodo della costruzione di un'etica dello scambio online, per la quale se vendevi un prodotto truffaldino non potevi più partecipare a un altro scambio, perché la stessa *community* si faceva garante e guardiana. Presto però ci si rese conto che pur senza mediazione, una forma di arbitrato andava quantomeno garantito e al posto dei mediatori apparvero i *selezionatori,* che organizzavano le collezioni su una piattaforma; i *curatori* che si assumevano il rischio di una scelta personale e solitamente tenevano un *blog;* gli *influencer* che si eleggevano a *guru* da seguire con fede indiscussa.

Nei fatti il web ci ha avvicinato a quanto desideriamo, ci ha messo a disposizione una pletora di possibilità, basta allungare la mano e prenderle. In mezzo non c'è più nessuna vecchia autorità, i passaggi si sono accorciati.

Si arriva da chiunque e ovunque. Le persone celebri, le straordinarie mete di vacanza, le incredibili avventure di chiunque, ci sono vicine se qualcuno le ha *postate,* le ha *tweettate,* le racconta ogni giorno come si trattasse dei nostri familiari: basta restare connessi.

Quindi il ruolo richiesto al designer non è quello di generatore di idee, ma di selezionatore e filtro di soluzioni possibili e migliori per il proprio committente, con un quoziente di legittimazione decisamente minore di prima.

La chiamavano *"reputation"*

Nel momento in cui si estinguono gli arbitri, allora il nostro valore, quello dei nostri lavori è dato dalla reputazione. Questa potrebbe funzionare come nel passato, su base meritocratica, basata sul giudizio positivo del mercato, dei clienti, degli *users*, degli abitanti.

Uno spazio è giudicato un buon progetto perché ad esempio, le persone lo visitano volentieri, oppure semplicemente perché il giudizio delle persone che ci sono state, è positivo.

Ammettiamo pure che ci sia un sistema di espressione di questo giudizio e che questo sia pubblico, allora quel giudizio potrebbe diventare la misura del valore del progetto stesso.

Raccontato in questo modo, viene da pensare che allora il web sia il migliore ufficio di comunicazione per chiunque abbia delle qualità da vendere o semplicemente esibire.

Inizialmente era così, ossia, quello che aveva molti *like* schizzava in alto nelle graduatorie della visibilità e se non ne aveva, o ne aveva pochi, precipitava nelle pagine più lontane. La rete sembrava avesse messo in atto una sorta di valutazione orizzontale della qualità, che faceva sperare in un mondo trasparente. Nel corso degli anni tuttavia si è sviluppata una forma di "economia della reputazione". Questa è divenuta moneta di scambio e, veri o presunti *followers* o utenti, esprimono dei giudizi sui progetti, prodotti, servizi, orientando il mercato semplicemente perché vengono pagati per farlo.

Per i siti di *hospitality*, o anche di semplice ristorazione, questo ha significato esaltare o demolire interi business, dirottando intere migrazioni di clienti verso un hotel o un altro, un ristorante o un suo concorrente, esclusivamente sulla base di recensioni non necessariamente veritiere.

Nella più sana delle condizioni, connaturata alle modalità del web, la *reputation* la creano i giudizi dei consumatori, dei clienti, degli *user*... e viene considerata più attendibile di quella degli esperti, mono di quella degli *influencer*. Questi ultimi, veri e propri *guru* della rete, sono in grado di spostare bacini di *like* ad uso e consumo del mercato e dei brand. Senza alcuna oggettività nel giudizio, né spesso competenza (se non come super-*users*), attraverso i social media maneggiano la potente, quanto pericolosa, macchina della reputazione.

Per garantire l'oggettività e accreditare i giudizi però dovrebbe esserci un sistema di controllo, di valutazione, di fiducia, quantomeno dei selezionatori. Dovrebbe esistere un sistema per il quale:

- i selezionatori sono sempre in buona fede e non presentano alcun conflitto d'interesse;

- gli algoritmi di selezione in rete, che rendono visibile il personaggio, l'immagine che ha più "like", venga considerata come quella che piace di più e non necessariamente come la migliore. Il che significa che il "parere di milioni di incompetenti è più affidabile, se sei in grado di leggerlo, di quello di un esperto";

- quello che va bene a tutti deve andare bene anche a noi, altrimenti la fatica di reperire in rete proposte diverse diventa una sorta di sfida a tutti gli algoritmi, che ci rimbalzeranno sempre verso la scelta più votata.

Sharing economy

Un altro paradigma profondamente cambiato nella e per via dell'era ditale è l'idea di proprietà. E' mutato il nostro rapporto con le cose e i luoghi che possediamo. Il valore della proprietà non consiste più esclusivamente nel detenerla, ma nel metterla in circolazione dentro un sistema di condivisione. L'idea è che se possiedi un bene questo è "mobile", se non lo stai usando tu, lo può usare qualcuno e pagarti il servizio.

La circolarità nell'utenza, che sia di un'auto, di una casa o di un ufficio genera un'economia della condivisione e produce un salto quantico nella idea degli spazi e del modo di progettarli.

Chi oggi chiede di progettare un'abitazione, non lo fa secondo una proiezione familiare, ma secondo un principio di business possibili. L'idea che intere zone dell'abitazione rimangano inutilizzate, perché un figlio si è trasferito all'estero per studiare e torna solo una settimana all'anno, oppure che durante le vacanze, l'intera casa rimanga "libera", oggi è considerato uno spreco. Per ragioni economiche, culturali e ambientali, questi beni son "messi in circolo".

Gli spazi residenziali, produttivi, terziari sono stati pienamente coinvolti nella possibilità che la *sharing-economy* gli ha offerto a sua volta, questa è stata recepita come *input* dal design degli spazi.

Nel momento in cui lo spazio non risponde più a un principio esclusivo di proprietà, ma ad una logica di rendimento, allora viene progettato su concetti diversi, con orizzonti temporali differenti, ma soprattutto il layout risponde a logiche di estrema versatilità e flessibilità. Oggi progettare uno spazio significa anche progettare le mille possibili vite che può raccontare.

Scena: Amazon nel 2017 ha lanciato un bando per la ricerca di una città che ospitasse la seconda sede del colosso (HQ2). Sono stati offerti 50mila posti di lavoro e stipendi in media di 100mila dollari all'anno. Ma i requisiti erano assai rigidi: un'area metropolitana con più di un milione di abitanti; un'area in grado di reclutare esperti specializzati; una distanza massima di 45 minuti da un aeroporto internazionale; un paio di chilometri di distanza da uno snodo autostradale; raggiungibile con trasporti pubblici.

Alla competizione, che è stata vinta da due città (New York e Crystal City in Virginia) hanno partecipato 238 città, alcune proponendo rilevanti tagli alle tasse del colosso, qualora la scelta le avesse favorite, altre proponendo addirittura di modificare il proprio nome con evidenti riferimenti al brand (Calmazon o Amalgary). Questo modo di intervenire sulle città, per nulla nuovo negli Stati Uniti, racconta un mondo in cui lo spazio fisico sarà valutato per le sue performance digitali almeno quanto quelle infrastrutturali.

Nuove forme di città e di spazi pubblici nasceranno da queste alleanze, saranno forse simili agli uffici di queste aziende: playground per adulti, in grado di trasformare il posto di lavoro, nel luogo da abitare, dove mangiare, dormire, giocare a ping-pong e trovare una compagna o compagno che necessariamente abbia la medesima vita da condividere. Saranno forse città commisurate a nuove forme di cittadini, nativi digitali, privi della smania di proprietà e rivolti a sistemi di reciproco scambio ed economie del tempo. Chissà, forse neppure useranno le automobili e lasceranno ai droni le consegne della merce e del pranzo o della cena. Avranno app per condividere e non perdere tempo a cercare, saranno belli e sgualciti. Non possiamo dire se saranno felici, ma vivranno più a lungo di noi, sapendo dallo *screening* del DNA di cosa andranno a morire.

Cercasi committente speciale

Nel magma generale che muove in continuazione tutti i pezzi del grande gioco del design, anche il cliente è cambiato. Quello che rimane un punto certo e fermo, almeno da qualche migliaio di anni, è che se il progettista non ha un buon cliente, il progetto sarà figlio di sofferenze e compromessi estenuanti che finiranno per penalizzare progetto, progettista e cliente.

Un cliente è buono quando è un buon committente, ossia quando sa scegliere il suo progettista con grande attenzione curriculare ed economica e poi sa lasciarlo lavorare senza troppe interferenze.

E' il committente a fare la differenza, almeno quanto il designer.

Cito un articolo pubblicato nel 1980 da Pierre-Alain Croset, in un bellissimo numero di "Rassegna" dal titolo *I clienti di Le Corbusier che* "egli ha lavorato soprattutto per dei clienti che gli offrivano l'opportunità di una sperimentazione, o che l'invitavano a costruire una "architettura dimostrativa". E se è possibile riconoscere che una vera e propria etica della sperimentazione guidò tutta la sua vita artistica, non si dovrà tuttavia dimenticare che nei margini del suo lavoro d'architetto può leggersi la continua *recerche patiente* del cliente ideale". Si capisce che quel genio di Le Corbusier, senza i suoi coraggiosi committenti non sarebbe andato lontano. Ad avvalorare questa tesi bisogna leggere un libro molto esplicativo delle dinamiche tra *Architettura e potere* di Deyan Sudjic in cui indaga il rapporto indissolubile tra potere e trasformazione degli spazi.

Il mondo è pieno di buoni progettisti, castrati da clienti incompetenti o ignoranti, così com'è pieno di celebri progettisti mediocri portati alle stelle da geniali clienti e riviste compiacenti. E su questo c'è poco da intervenire.

Cosa vuole il cliente

Quando il committente chiama un designer lo fa perché:

- non sa bene cosa vuole, ma sa cosa non vuole, e gli chiede di aiutarlo a porsi la domanda più corretta per sé: chiede una STRATEGIA

- ha un problema e vuole che una persona esperta glielo risolva. In questo caso il designer, come il dottore, deve trovare una cura, deve agire da *problem solver*, in maniera lucida e tecnica: vuole un SERVIZIO

- aspira a guardare il mondo con le lenti del designer che ha scelto, vuole che creino qualcosa che lui da solo non è in grado di progettare o semplicemente di esprimere: chiede un atto di CREAZIONE.

Esiste un grande fraintendimento, che coinvolge tutti gli aspiranti designer: che la progettazione sia solo un atto creativo. E' indubbiamente uno dei primi traumi del giovane laureato, scoprire che non è propriamente così che va il mondo. La creatività è una parte il resto è competenza professionale (progettuale, tecnica, commerciale, relazionale, culturale...). Questo concetto semplicissimo, e quasi banale, è una delle fondamentali cause di frustrazione di moltissimi designers del pianeta.

Il mercato richiede di interpretare, di rispondere, di progettare significati, o concatenazioni di significati in forma di racconti (*story telling*). La creatività è quel *plus*, che ha un valore inestimabile, che è la cifra con cui ogni designer, davanti al medesimo problema, trova la propria risposta migliore.

Nei confronti della rivoluzione digitale in corso, bisogna abbandonare qualsiasi forma di nostalgia e provare a compiere la più radicale mutazione genetica che il design degli spazi abbia subìto sin dalle sue origini: quello di rinunciare all'autorialità e modificare i suoi romantici paradigmi.

La paura dell'innovazione

Uno dei miti del XX secolo è quello che il mondo del design debba sempre ricercare l'innovazione e che il successo sia legato al conseguimento della soluzione più innovativa possibile.

Questo mito ha raggiunto picchi di fanatismo, nel design di prodotto, ai tempi delle grandi visioni di Steve Jobs, in grado di trascinare anche i comuni mortali dentro mondi mai visti prima.

Quelle visioni appartenevano ad un'era che definirei di "corsa all'ultima invenzione" che non è sempre *main stream*. Pensiamo agli *smart phone*, il 2017 è stato l'anno in cui le vendite sono per la prima volta diminuite. In quel settore il mercato è maturo, saturo, e il costo è proporzionalmente alto, tanto quello di un'auto di lusso. In futuro ci accontenteremo di comprare anche qualche modello precedente, perché la sensibilità all'innovazione è già significativamente appagata.

Non solo il mercato ma anche i designer, pensano che progettare richieda sempre un quoziente indispensabile di novità. Ci comportiamo come degli stilisti alla rincorsa di una qualche tendenza da seguire, di qualche gusto insolito da soddisfare. Ecco non è così che funziona.

Nella vita di un designer è chiaro da subito che non tutti i clienti sono uguali e che alcuni clienti non hanno bisogno, e/o intenzione, e/o coraggio, e/o necessità di lanciarsi in avventure inesplorate.

Ci chiedono anzi di progettare qualcosa che assomigli a un'altra, semplicemente di declinarla nel loro contesto.

Il paradosso del rendering

Una società così evoluta in termini d'immagini, usa la vista per comunicare e conseguentemente richiede al design modelli verosimili e realistici degli spazi, in grado di raccontarli come se già esistessero.

L'astrazione tipica dello schizzo, viene bandita – soprattutto nel design degli interni- a favore di rendering/video tanto accurati, quanto prematuri, richiesti dal cliente praticamente a pochi giorni dall'avvio del progetto, per "vedere come sarà" per decidere se gli piace o meno.

Come se un cliente al ristorante chiedesse di assaggiare la pasta appena qualche minuto dopo averla messa in cottura... e continuasse a chiedere in continuazione, con il rischio e che al momento di doverla servire nel piatto ne sia rimasta poca o nulla e con la certezza che il cliente, che l'ha assaggiata cruda durante tutta la fase di cottura, poi si lamenti che non gli piace o addirittura che sia ancora cruda.

Quest'ansia di visualizzazione dello spazio nelle sue qualità finali è uno dei problemi maggiori che affronta il designer, che deve rispondere a una richiesta del committente che si comporta con il progetto di spazio come se fossero articoli da comprare *online*, già disponibili.

I rendering sono forzati verso un modello ribaltato nell'elaborazione di un progetto: partono dall'immagine e da essa provano a estrapolarne dei luoghi (nei migliori dei casi). Oggi è anacronistico opporsi al desiderio di vedere e i rendering sono un media efficace per un racconto verosimile degli spazi.

Oggi, a torto o a ragione, il committente vuole vedere il suo progetto subito, per decidere se conferire l'incarico professionale. Non vuole fare lo sforzo di traduzione nel suo contesto delle idee del designer, quindi chiede di vedere il progetto prima ancora che sia stato progettato o addirittura pensato.

Il paradosso è di produrre un render o un video, il più verosimile possibile, ma completamente arbitrario nelle sue scelte, sin da subito, affinché riceva approvazione dal cliente, salvo poi rimanerne vittima, perché ad ogni ripensamento e variante, il vostro cliente ne chiederà conto.

I siti web di studi di design sono popolati da immagini di rendering talmente verosimili da non distinguerli più da opere costruite, rendendo impossibile, a occhio inesperto, distinguere uno studio di progettazione da una *render farm*.

Esercizi di fotogenìa

Così come noi progettisti accediamo alle *libraries* di immagini *online*, anche il mercato le consulta, le sfoglia, si chiarisce cosa gli serve o cosa desidera. Le piattaforme online che pubblicano progetti, potrebbero essere quindi luoghi di incontro tra designers e clienti. Invece diventano spesso banali *mall* virtuali per shopping online di aspirazioni, senza capire che scegliere una immagine di uno spazio è diverso da immaginare uno spazio e poi abitarlo.

Ceci n'est pas une pipe intitolava Magritte un'opera nel 1926, che divenne poi esercizio seriale, sul rapporto tra rappresentazione, descrizione e realtà. Ed è proprio questa la pretesa più frequente di comprendere gli spazi attraverso le immagini.

Copiare immagini non significa progettare dei luoghi, attività che comporta invece attenzione alle luci (naturali e artificiali); ai suoni presenti e a quelli che si generano, agli odori, alla temperatura, al *touch* manuale delle superfici e al *touch visuale* delle finiture, al racconto cromatico, all'umidità dell'aria, alla presenza di altri umani...

Abituati alla produzione sempre più sofisticata di immagini, anche il design degli spazi ne subisce il fascino, trascurando, talvolta, la partecipazione degli altri sensi alla progettazione della bellezza e riducendo il design degli spazi a un mero esercizio di fotogenìa.

Instagrammabilità

Fino a qualche anno fa, andare a vedere un luogo bello comportava anche scattare foto ricordo e portarsele con sé. Oggi in quel medesimo rituale, lo scatto prevede che il fotografo sia anche il protagonista e che il luogo sia invece sia la quinta, lo sfondo. L'era dei *selfies* non è una novità, ma questa singolare maniera di rapportarsi a qualsiasi luogo come a uno scenografia, pone degli interessanti risvolti sul design degli spazi.

Alcuni musei ormai sono "selfiefici" dove le opere servono esclusivamente a testimoniare la presenza effettiva del visitatore in quel sito specifico. La visita in sé è meno importante di quello scatto da "postare" più o meno immediatamente. Il Victoria & Albert Museum ha addirittura introdotto di recente l'*hashtag* #myvam, in modo da ottimizzare l'abitudine dei suoi visitatori e usa le loro foto sui *social* come comunicazione e promozione del museo stesso.

Instagram e i social media condizionano i nostri progetti, perché la loro vita social ha un impatto sul loro successo pari almeno alla qualità della progettazione. Le aziende sul mercato chiedono ai progettisti un quoziente di "instagrammabilità" sempre più significativo. I selfie-points diventano dei *place-mark* che devono diventare icone social dei brand per farne pubblicità.

Il Rove Hotel di Dubai, della società Stride Treglowe, ha deciso di mettere *hashtag* ovunque, perché il target dell'albergo è *social*-mente attivo a partire dall'ingresso: #*this is where I am*.

Sono progettati con la medesima logica, il negozio di Nike a Shanghai, dove l'agenzia di *branding* Rosie Lee ha costruito un trono di scarpe e i magazzini Harrod's di Farshid Moussavi, che arriva addirittura a dichiarare che "l'instagrammabilità è ormai parte del briefing, specialmente nel mondo dell'ospitalità".

Il *Cloud Gate* di Anish Kapoor a Chicago è tra i *selfie-points* più postato al mondo. Via via nascono luoghi la cui unica funzione è quella di essere icone.

La richiesta esplicita dell'imprenditore che ha commissionato, per 200 milioni di dollari, il Vessel di Thomas Heatherwick a New York, è stata infatti quella di avere "una Tour Eiffel per turisti". Il progetto, proprio come la Tour Eiffel, è una teorie di scale in un sistema piranesiano e parametrico che non portano a nulla, se non alla esperienza del salire... del fotografarsi... del "postare".

Il design per gli spazi da pubblicare sui social comporta delle nuove armonie, delle nuove regole, dei nuovi canoni di una proporzione post-vitruviana: quelle quadrate di *instagram*. Lo studio Vale Architects ha pubblicato di recente una *Instagram Design Guide*, affinchè i progettisti rendano gli interni più "instagrammabili" come piace al mercato.

Parte 3 _ #sensi

L'uomo del 2598 (Bruno Munari)

Uno studioso del futuro (il signor Jean-Pierre Esposito Tung, nato nel 2598) ha ricostruito, per esempio, il tipo umano vivente nel periodo che va dal 1925 al 1971, in base alla qualità e alla varietà delle poltrone, sedie divani costruite in quell'epoca. Pare che, soprattutto per l'accanimento con il quale questo problema veniva affrontato e progettato, come fosse il più importante di tutti gli altri, il tipo umano vivente in quel periodo dovesse avere un sedere enorme, gonfiato dalla mancata traspirazione a causa delle arterie plastiche usate per costruire o rivestire sedie, divani e poltrone... E così si progettano bellissime sedie rumorose quando si spostano, bellissime poltrone che pesano un quintale o un etto; gli ambienti in cui non circola l'aria. Nelle case di progettazione non si tiene conto dell'odore, come se il corpo fosse sprovvisto di naso, per cui odori sgradevoli circolano liberamente dappertutto, il fritto di pesce va a finire nell'armadio e la naftalina in cucina. Il mondo del rumore è addirittura ignorato: scarichi di wc in soggiorno, rumori di ogni tipo da ogni parte; senza parlare della progettazione di ristoranti bellissimi e, di lusso, dove il rumore è assordante. Se il designer è il coordinatore di tutte le componenti di un progetto, dipenderà anche da lui se l'uomo conserverà o atrofizzerà alcuni sensi. Si sa che come la funzione sviluppa l'organo, la non funziona lo atrofizzerà, molti animali che vivono in ambienti diversi rispetto alla loro origine perdono l'uso di certi organi, ad esempio animali diventati ciechi perché vissuti molto tempo in grotte buie, così anche l'uomo può perdere l'uso di certi organi se essi non sono mantenuti stimolati. Bisogna ricordarsi che l'individuo è dotato di tutti i sensi e che esiste anche un piacere di toccare materie gradevoli, che il legno è più gradevole al tatto del metallo, che certe materie lasciano respirare il corpo e altre no, che i rumori danno fastidio anche quando ci si è abituati, che gli odori non sempre sono piacevoli, che l'acuto stridore di certe carrozzine per bambini rovina le orecchie ai neonati... Se, come pare, la funzione sviluppa l'organo; la non funzione lo atrofizzerà. Vedremo quindi nel futuro uomini senza orecchie? O senza naso? O con la schiena e il sedere deformati dalla mancata traspirazione? Sarà questo l'uomo del futuro? Speriamo di no. Ricordiamoci quindi, quando progettiamo qualcosa che abbia anche un buon senso tattile la gente se ne accorgerà e ricomincerà a usare di nuovo questo che è uno dei più trascurati dei sensi. Se teniamo conto anche degli altri sensi, la gente pian piano si abituerà e scoprirà che ci sono tanti recettori sensoriali per conoscere il mondo in cui viviamo. I bambini lo sanno benissimo e la prima conoscenza del mondo è, per loro, sensoriale e globale...

Sensefulness

Lo spazio digitale è pulito, asettico, privo di attrito, di peso, di consistenza...
è un mondo audio-visuale in cui abitare superando i limiti fisici del mondo
gravitazionale.

Ci siamo talmente abituati alla velocità del web che l'idea di stare in coda,
di aspettare non è più accettabile. L'ebbrezza della rete crea dipendenza.
Il mio corpo digitale sono le informazioni su di me che stanno in rete, a
discapito della verità e della coerenza... e del mio corpo.

Quando partecipiamo al mondo digitalmente, il corpo assume delle
posture diverse, seduto, sdraiato, ormai itinerante con una mano occupata
da quella protesi digitale che è lo *smart-phone*.

Il corpo durante l'interazione digitale è spesso privo di consapevolezza
spaziale e sensoriale dello spazio che lo circonda. Niente attrito, niente
freddo, niente peso, niente puzza. Il centro del racconto è dentro la rete,
sebbene il fronte delle emozioni sia sul corpo e dentro di esso.

Che lo spazio sia buio, rumoroso, illuminato... il suo impatto è minimo
perché il corpo è emotivamente eccitato più per quanto accade aldilà
dello schermo che non aldiquà. Dal lato della realtà c'è un pachiderma
spiaggiato, dal lato della virtualità c'è un'atletica gazzella che corre in un
mondo che teatralizza ogni rappresentazione.

Il rapporto causa-effetto tra i nostri movimenti è esponenziale: a pochi gesti
corrispondono gesti infiniti.

Anche mentre viviamo nella rete, il nostro corpo è in uno spazio fisico, il
corpo c'è e le emozioni –anche quella suscitate dalla rete- sul corpo
ricadono.

Sembra una sentenza assai banale, ma è invece cardinale, perché
sebbene le nuove tecnologie continuino a essere ocularcentriche,
l'esperienza dei luoghi, ricerca anche negli altri sensi un completamento
emotivo irrinunciabile.

Se chiedete alla gente normale (non ai designer quindi) di raccontarvi
un luogo, il suo racconto andrà oltre lo sguardo: vi parlerà del silenzio,
dell'affollamento, del caldo o del freddo, degli odori che ha sentito o
del confort e via così con una serie di descrizioni che restituiscono la sua
esperienza di quel luogo: un'alchimia tra qualità oggettive e sensazioni
personali che costituiranno non solo la sua esperienza, ma il ricordo che
avrà di essa.

La questione centrale non è più quella di attaccare identità sulle cose,
ma di progettare quella che ci sta dentro, nel loro DNA. L'identità non è
un'etichetta che si appone successivamente. Le identità si vivono, non
s'indossano.

Ripartire dall'esperienza dei luoghi, restituisce gli spazi alla gente comune, gli conferisce il ruolo di attori e non di spettatori di quello spazio che abitano tutti i giorni e che possono/devono modificare.

Così, alla "rivoluzione digitale" che ha mutato gli spazi, ne ha scomposto le geometrie e modificato gli strumenti di misurazione è subentrata la "rivoluzione temporale". Questa ha comportato la progettazione di forme di tempo piuttosto che forme di spazio, ha iniziato l'era della economia della condivisione e ha posto nuove questioni ancora aperte. Adesso è il momento di una "rivoluzione sensoriale", in cui i sensi diventano media fondamentale nella integrazione tra reale e virtuale, in cui il corpo rivendica il suo ruolo di terminale delle emozioni sia che provengano dalla realtà, sia dalla virtualità.

Sense_based design

Qualche mese fa mi è arrivato un *briefing* per un progetto di *Spatial Design* da realizzare in un paese straniero. Per la prima volta nella mia vita professionale, il *briefing* comprendeva uno schema molto interessante in cui era inserita una tabella di requisiti richiesti per i singoli spazi. Si richiedeva espressamente di progettare l'esperienza dei luoghi e in maniera sistematica, si elencavano le esigenze secondo quattro categorie: umana, sensoriale, comunicazione, fisica.

Ecco, ho pensato, quello era il segno che qualcosa stava andando a recuperare una necessità, la relazione persa con i luoghi, attraverso il corpo, i sensi, l'esperienza umana e quella fisica.

Il *briefing* molto dettagliato indicava la necessità di avere spazi illuminati con luce naturale, di avere la possibilità di assumere diverse posizioni del corpo durante il giorno, di sperimentare delle posture nuove. Si richiedeva una sensibilità acustica, non solo per attutire i rumori, ma tramutare le varie aree in ambienti sonori dove poter parlare senza essere sovrastati dalla musica indesiderata. Si richiedevano superfici che invitassero al contatto, che non fossero troppo riflettenti o lucide e che giocassero con la luce un sottile design di modellature. Chiedeva che l'aria non fosse troppo "condizionata", anzi invitava a lasciare circolare alcuni odori che dessero identità al luogo e magari poter aprire qualche finestra. Che ci fosse un orientamento temporale, che aiutasse in qualsiasi momento le persone a capire che ore fossero, senza dover guardare l'orologio. E che di notte il medesimo edificio continuasse a vivere, magari un'altra vita lasciando al progettista l'onore di progettarne più di una.

Era da un pò di anni che mi aspettavo che questo accadesse di nuovo, che si smettesse di enfatizzare solo la dimensione tecnologica o le qualità comunicative degli spazi. Me lo sentivo nel profondo che il corpo si sarebbe ribellato e avrebbe voluto riprendersi la scena nello spazio che lo ha visto protagonista per millenni. E' un corpo diverso ed è uno spazio riformato, ma la necessità di riprendere la dimensione fisica è tornata. Il corpo rimane la sede delle nostre emozioni, che provengano dalla camera accanto o dalla rete, è sempre lui che reagisce.

Attraverso i sensi, si ricostruisce anche l'empatia, quella capacità umana e animale di sentirsi un tutt'uno con gli altri, di immedesimarsi nelle loro sensazioni. Davanti a uno schermo, nell'interazione con le macchine, la comunicazione empatica lascia il posto alla comunicazione efficiente e si perde la capacità di comprendere appieno chi ci sta di fronte.

"Il nostro inconscio sta diventando un raffinato maestro nello sfuggire all'attesa, alla comprensione e all'attenzione dedicata, ovvero alla triade della comunicazione empatica" scriveva il *digital strategist* Giulio Xhaet su "Il Sole 24 Ore" del 16 ottobre 2018.

Solo progettare con la mediazione dei sensi, può guidare l'immaginazione, verso le altre dimensioni percettive e conseguentemente emotive ed empatiche.

Mai come oggi il kit del progettista, che serve per questa nuova spedizione, è una cassetta di attrezzi sensoriali, in grado di aiutare a interpretare i segnali di questa ennesima rivoluzione e soprattutto di incoraggiare una serie di linguaggi e pratiche, già in corso, che ci portano a una maggiore condivisione e consapevolezza degli spazi che abitiamo.

Mi piacerebbe chiamare questa pratica del design, *sensefulness*, per ottenere una esperienza immersiva dei luoghi senza bisogno di troppe allegorie.

Così come mi piacerebbe chiamare *sense_based design* la disciplina che usa tutti i sensi per progettare luoghi, prodotti, servizi, brand che favoriscano esperienze consapevoli attraverso il corpo e i suoi apparati.

La risposta alla sovraeccitazione visuale è una progettazione completamente sensoriale, in cui i sensi non intervengano come decorazione fittizia alla fine del processo di progettazione, ma che siano invece strumenti costituenti il DNA del design e dei suoi spazi.

Bibliography

Augé M., *Che fine ha fatto il futuro?*, Elèuthera, 2009

Augé M., *Nonluoghi: introduzione a una antropologia della surmodernità*, Milano, Eleuthera, 1993

Barbara A., Storie di architettura attraverso i sensi, Postmedia Books, Milano, 2011

Barbara A., *Sensi, tempo e architettura*, Postmedia Books, Milano, 2012

Barbara A., Ceresoli J., Chiodo S., *Interni inclusivi*, Maggioli Editore, 2016

Baricco A., *The Game*, Einaudi, Torino, 2018

Bauman Z., *Modernità liquida*, Laterza, Roma-Bari, 2002

Bauman Z., *Consumo, dunque sono*, Editori Laterza, 2007

Birnbaum D., *Cronologia*, Postmedia Books, Milano, 2007

Certeau M. de, *L'invenzione del quotidiano*, Edizioni Lavoro, Roma, 2005

Danzi O., Re G., *Community manager. Dietro le reti ci sono le persone*, Franco Angeli, Milano 2018

Faré I., *Il discorso dei luoghi: genesi e avventure dell'ordine moderno*, Liguori, Napoli, 1992

Faré I., Cascitelli L., Barbara A., Bassanini S., Piardi S., *Nuove specie di spazi*, Liguori editore, Milano, 2003

Gensler A., *Art's Principles*, Wilson Lafferty, US, 2015

Giedion S., *Breviario di architettura*, Bollati Boringhieri,Torino, 2008

Giedion S., *Space, Time and Architecture: The Growth of a New Tradition*, Cambridge/Mass, 1967

Gwiazdzinski L., *La Ville 24 heures sur 24*, editions de l'aube datar, 2003

Heidegger M., *Il concetto di tempo*, Adelphi, Milano, 1998

Heidegger M., *Essere e tempo*, UTET,Torino, 1986

Jung C.G., *La sincronicità*, Bollati Boringhieri,Torino, 1980

Mau B., *Cinematic Migration*, Phaidon; London, 2000

Merleau-Ponty M., *Fenomenologia della percezione*, Bompiani, Milano, 2003

MVRDV/UWM, *Skycar City. A pre- emptive History*, Actar, 2007

Norberg-Schulz C., *Genius Loci: paesaggio, ambiente, architettura*; Electa, Milano, 1979

Rykwert J., *La seduzione del luogo. Storia e futuro della città*, Einaudi, Torino, 2003

Thackara J., *Progettare oggi il mondo di domani*, Postmedia Books, Milano, 2017

Touraine A., *La società postindustriale*, Il Mulino, Bologna, 1970

Verganti R., *Overcrowded, Il manifesto di un nuovo modo di guardare all'innovazione*, Hoepli, Milano, 2016

Virilio P., *La bomba informatica*, Cortina, Milano, 2000

Virilio P., *Polar Inertia*, Sage, Londra, 2000

Sensefulness
New paradigms for Spatial Design
Nuovi paradigmi per lo Spatial Design

Anna Barbara

postmedia books 2018
128 pp.
ISBN 9788874902378

Published by Sartoria editoriale Milano 2020

Postmedia Srl
Milano

www.postmediabooks.it

www.ingramcontent.com/pod-product-compliance
Lightning Source LLC
LaVergne TN
LVHW060345200726
843507LV00005B/976